# **Algebra**

By Robert Watchman

An Easy Steps Math series book

Copyright © 2014 Robert Watchman

All rights reserved.

No portion of this publication may be reproduced, transmitted or broadcast in whole or in part or in any way without the written permission of the author.

## Books in the Easy Steps Math series

Fractions
Decimals
Percentages
Ratios
Negative Numbers
Algebra
Master Collection 1 – Fractions, Decimals and Percentages
Master Collection 2 – Fractions, Decimals and Ratios
Master Collection 3 – Fractions, Percentages and Ratios
Master Collection 4 – Decimals, Percentages and Ratios

More to Follow

# Contents

| | |
|---|---|
| Introduction | 7 |
| Chapter 1 **Algebra Basics** | 9 |
| Chapter 2 **Algebraic Language** | 11 |
| Chapter 3 **Like Terms** | 14 |
| Chapter 4 **Substituting Numbers** | 18 |
| Chapter 5 **Exponents (Powers) in Algebra** | 24 |
| Chapter 6 **Multiplying Algebraic Terms** | 30 |
| Chapter 7 **Dividing Algebraic Terms** | 33 |
| Chapter 8 **Multiplying and Simplifying Algebraic Expressions** | 36 |
| Chapter 9 **Factoring (Factorising) Algebraic Expressions** | 49 |
| Chapter 10 **Algebraic Fractions** | 76 |
| Chapter 11 **Solving Linear Equations** | 88 |
| **Multiplication Tables** | 100 |
| **Answers** | 102 |
| **Glossary of Useful Terms** | 108 |

# Introduction

This series of books has been written for the purpose of simplifying mathematical concepts that many students (and parents) find difficult. The explanations in many textbooks and on the Internet are often confusing and bogged down with terminology. This book has been written in a step-by-step 'verbal' style, meaning, the instructions are what would be said to students in class to explain the concepts in an easy to understand way.

Students are taught how to do their work in class, but when they get home, many do not necessarily recall how to answer the questions they learned about earlier that day. All they see are numbers in their books with no easy-to-follow explanation of what to do. This is a very common problem, especially when new concepts are being taught.

For over twenty years I have been writing math notes on the board for students to copy into a note book (separate from their work book), so when they go home they will still know how the questions are supposed to be answered. The excuse of not understanding or forgetting how to do the work is becoming a thing of the past. Many students have commented that when they read over these notes, either for completing homework or studying for a test or exam, they hear my voice going through the explanations again.

Once students start seeing success, they start to enjoy math rather than dread it. Students have found much success in using the notes from class to aid them in their study. In fact students from other classes have been seen using photocopies of the notes given in my classes. In one instance a parent found my math notes so easy to follow that he copied them to use in teaching his students in his school.

You will find this step-by-step method of learning easier to follow than traditional styles of explanation. With questions included throughout, you will gain practice along with a newfound understanding of how to complete your calculations. Answers are included at the end.

# Chapter 1

## Algebra Basics

Too many people find algebra difficult. Too many people get worried when the idea of algebra comes up. Even students, who have not yet started to learn algebra, but have heard that it is difficult, want to give up before they start. This tends to be the case if the work is not explained well in the first instance, or if the student wasn't paying attention when it was explained, and the teacher did not go back or could not go back to explain it again as thoroughly as the first time, if at all.

Many students are too busy telling themselves how hard algebra is going to be rather than listening to how easy it actually is. Having said this, many students have learned the basics of algebra from elementary school or primary school and did not know it. For instance, students can easily answer a questions like, *solve* ___ + 3 = 5. The answer is obviously 2. Or a question like *solve* 3 × ___ = 15 can also be worked out easily, and the answer is 5. This answer is usually written into the space that has the underscore. These questions are the basics of algebra. An underscore, or sometimes a box, is put in, in the place of a letter. To rewrite these questions algebraically would make them look like this:

___ + 3 = 5 becomes $x + 3 = 5$. The answer you would write would be $x = 2$.

3 × ___ = 15 becomes 3 × $x$ = 15. The answer is $x = 5$.

Notice how the letter $x$ is written differently to the multiplication sign ×. It is very important that you do this because it gets very confusing if they were both written the same way. Many students who don't pay attention to this get very mixed up with their work and get many answers wrong and don't understand why. They write ××××× instead of $xxxxx$.

When multiplying in algebra, it is not necessary to write in the multiplication sign all the time. When letters (or variables) are written next to each other, as in $ab$, this has the same meaning as $a \times b$. If there is a number written in front of these as in $5ab$, this has the same meaning as $5 \times a \times b$. However, if there is nothing written in front of a variable, like $x$, or $ab$ then this is the same as $1x$ which is $1 \times x$, or $1ab$ which is $1 \times a \times b$.

You would know from the Fractions book that a fraction is just a division, so when you see $a \div b$, this is the same as $\dfrac{a}{b}$. When you see an expression like $5 \times a \div b$, this is the same as $\dfrac{5a}{b}$.

Generally in algebra the number is written in front of the letter like in $4x$. If a question asked you to multiply the letter $a$ with the number 10, your answer would be written as $10a$ and not $a10$.

**Rewrite the following without × and ÷ signs.**

a) $x \div 6$

b) $3 \div y$

c) $4 \times a$

d) $b \times 7$

e) $5 \div (a \times 2)$

f) $9 \div a - x \div 7$

g) $a \times b \div c + 3 \times a$

h) $9 \div (3 \times s) + 6 \times r \div (u \times v)$

i) $4 \times e \times f \div (q \times t)$

j) $s \times v \div 7 + 3 \times 5$

# Chapter 2

## Algebraic Language

The first things to learn in algebra are some of the words used to describe the different parts. When you have questions that use some of these words, you need to be able to understand what the questions are asking otherwise there is no chance of being able to answer them. Here are some of the words you need to know.

**Variable**, or **pronumeral** – This refers to the letters or other symbols used in algebra. They are called variables because you can vary or change the numbers that you can substitute into them.

E.g. $x$ is a variable, so is $a$ and so is any other symbol or groups of letters and symbols such as $ab$ or $xy$ or $\eta\theta\sigma$ etc. The $x$, or any other variable can be any number that you give it. It can be a 3, or 12, or $-15$, etc. Sometimes the value of the variable is not known and this is what needs to be worked out.

**Coefficient** – This is the number in front of the variable.

E.g. In the term $3x$, the 3 is the coefficient. In the term $4x^2$, the 4 is the coefficient. In the term $y$, there is no number written in front of the variable, therefore the coefficient is 1. Remember that $1 \times y = y$.

**Exponent** or **power** – This is the small number written to the top right of a variable or number. It means the letter or symbol or number below it, called the base, is multiplied by itself the same number of times as the small number shows.

E.g. In the term $x^3$, the small 3 is the exponent and the $x$ is the base It means the pronumeral $x$ is multiplied by itself three times, that is

$x^3 = x \times x \times x$. In the term $a^4$, it means the base $a$ is multiplied by itself 4 times, that is $a^4 = a \times a \times a \times a$. If there is a number instead of a variable, like $3^3$, then this is the same as $3 \times 3 \times 3 = 27$. Where no exponent is written, the power is 1. Any number or variable with the power 1 stays the same so $3^1 = 3$ and $x^1 = x$. Exponents and powers will be covered in more detail later.

A **term** refers to a grouping of coefficients and/or variables that are multiplied or divided with each other.

E.g. In the expression $4x^2 + \frac{7x}{2} - 2$, the first term is $4x^2$, the second term is $\frac{7x}{2}$ and the third term, the $-2$, is called the constant. Each term is separated by a plus (+) sign or a minus (−) sign.

An **expression** is a group of two or more terms that are separated by a (+) or (−) operation sign. For example $4x^2 + \frac{7x}{2} - 2$ is an expression and so is $2x + y$. A **binomial expression** has 2 terms like $2x + y$ or $x^2 + 3y^3$, a **trinomial expression** has three terms like $4x^2 + \frac{7x}{2} - 2$ or $3r + 7s + 2t$. These expressions can also be referred to as polynomial expressions. The word **poly** meaning more than one or many.

An **equation** is where two expressions are equal to each other. An equals (=) sign is included to show that one side equals the other side. There may be two expressions that equal each other, or an expression is equal to a term or maybe a term is equal to a constant, or any combination of these.

E.g. $x^2 = 9$ is an equation because the expression $x^2$ has been made equal to the value 9.

$3x+2y=9$ is also an equation because the equals sign makes it so.

A **constant** is just a number in an expression. It is called a constant because its value does not change. In the expression $4x^2+\frac{7x}{2}-2$, the constant is $-2$. This number does not change no matter what values you put in for $x$. The value of the expression changes depending on what number you substitute into $x$, but the $-2$ will stay as $-2$.

You should remember from the Easy Steps Math Negative Numbers book that numbers are either negative or positive. Therefore the constant will be a negative value if there is a minus $(-)$ sign in front of it.

All expressions have a constant, but sometimes it is not written. This is because the constant can be a $0$ and you don't need to write this in.

**For each of the following expressions, identify the following parts:**

i) the variable of the first term  ii) the coefficient of the second term
iii) the exponent of the first term  iv) the second term  v) the constant

a) $a+b$

b) $a-b-2$

c) $3a+2b$

d) $x+y+3$

e) $8x+4y+2$

f) $2x+3y+4z$

g) $x^2+2y$

h) $4x^2+3y+2$

i) $x^2-25$

j) $3x^2y+\frac{x}{4}-2$

# Chapter 3

## Like Terms

Like terms are terms where the <u>variables are exactly the same</u>. The coefficients can be different and the order the letters are written in does not matter. If terms are like terms then you can add and/or subtract them.

For instance, the term $x^2y$ is exactly the same as $yx^2$. The variables are the same, but the order they are written is different.

Another example is, $3a^2b^3c$ and $2cb^3a^2$ are like terms. The variables are exactly the same, but the order they are written in is different and so are the coefficients.

Terms can look a bit busy sometimes so you may need to check each variable carefully before adding or subtracting like terms.

Like terms can be added and subtracted, but unlike terms cannot. In order to simplify an expression, you would 'collect' like terms, which refers to adding or subtracting them.

For example,
In the expression $3a+4b+5a+3b$ the $a$s are like terms and the $b$s are like terms. So to simplify this expression, adding the $a$s, you would get $8a$, and adding the $b$s, you would get $7b$.

$\therefore 3a+4b+5a+3b = 8a+7b$. This cannot be simplified any further.

In an expression like $4d+8e+5d-7e$, collecting like terms gives the answer of $9d+e$. Notice how $8e-7e$ gives the result of $e$, and not $1e$. As mentioned above if the coefficient of a variable is 1, you don't need to write it in. $1e$ is the same as $1 \times e$ which is the same as $e$.

In the expression $3a^2+4b+5a+3b$, the $a^2$ is **not** the same as the $a$, so they <u>are not</u> like terms. To simplify $3a^2+4b+5a+3b$, you would get $3a^2+5a+7b$.

The same applies to $x$s and $y$s and $z$s and any other variables or terms.

The term $x^2y$, is not the same as $xy$, so they are not like terms.

In the expression $3x^2+4x+2x^2+5xy+7x$, there are three different terms. Those with $x^2$ are like terms with each other only. Those with $x$ are like term with each other only and those with $xy$ are like terms with each other only. So to simplify the expression $3x^2+4x+2x^2+5xy+7x$ you would get $5x^2+11x+5xy$.

If you have a long expression, an easy way to reduce mistakes is to cross out the like terms you have added or subtracted. This ensures you don't add or subtract the same terms twice. Be careful with the positive and negative signs. If you're unsure about this refer to the Easy Steps Math Negative Numbers book.

For example

*Simplify* $6a^3-8a^2-4a-7a^3-5a-4a^3+a^2-61$

**Step 1.** Collect all the terms with $a^3$, then cross them out in the question.

You would get $\cancel{6a^3}-8a^2-4a\cancel{-7a^3}-5a\cancel{-4a^3}+a^2-61$

Which equals $-5a^3-8a^2-4a-5a+a^2-61$

**Step 2.** Collect all the terms with $a^2$, then cross them out in the question

You would get $-5a^3 \cancel{-8a^2} -4a-5a\cancel{+a^2} -61$

Which equals $-5a^3-7a^2-4a-5a-61$

**Step 3.** Collect all the terms with $a$, then cross them out in the question

You would get $-5a^3-7a^2\cancel{-4a}\cancel{-5a}-61$

Which equals $-5a^3-7a^2-9a-61$

As there are no other like terms, the constant stays as it is and the answer is $-5a^3-7a^2-9a-61$

**Simplify the following expressions**

a) $5a+12a$

b) $10f-3f$

c) $-7w+3w$

d) $4ef^2g^3+9g^3ef^2$

e) $8d-3e+4d-7e$

f) $8ij+12ij-4ij+3j$

g) $6x^3 - y^2 + 4x^3 + 9y^2$

h) $5x^2 - 6y - 13x^2 - 4y + 2$

i) $-9b - 4b^3 + ab - b^3 + 8b - 6ab + 8b^3$

j) $9s + 40st + 5 + 100stc - 5c + 15 - 2s + 50tcs$

# Chapter 4

## Substituting Numbers

In algebra, sometimes you will need to replace a variable with a number in order to solve it. This is called **substitution**. Substitution becomes important when having to simplify expressions, equations and formulae (plural for formula).

As mentioned earlier, the multiplication sign ($\times$) is not always shown in algebra. Therefore the term $3 \times a$ is written as $3a$. When substituting however, it is better to put the multiplication sign back in. This helps to avoid confusion. Consider the following. If $a=2$, then $3a$ would become $3 \times 2$ which equals 6. If you didn't put the multiplication sign in then $3a$ would look like 32, which would cause all sorts of problems.

Examples:

E.g. 1. The formula to convert degrees Celsius (C) to degrees Fahrenheit (F) is $F = \dfrac{9C}{5} + 32$. Use this formula to convert the temperature from 28°C to degrees Fahrenheit.

Step 1. Replace the C in the formula with the number 28 remember to put the $\times$ sign in.

$$F = \dfrac{9 \times 28}{5} + 32$$

Step 2. Following the Order of Operations, multiply the 9 and the 28.

$$F = \dfrac{252}{5} + 32$$

**Step 3.** Complete the division then the addition to get your answer.

$F = 50.4 + 32$

$F = 82.4$

So 28 degrees Celsius is the same as 82.4 degrees Fahrenheit.

E.g. 2. Sometimes you will have an equation like $y=3x+4$, and the question will ask you to find the value of $y$ when $x=5$.

Follow these steps:

**Step 1.** Replace the variable with the number. Remember to put the $\times$ sign in.

$y = 3 \times 5 + 4$

**Step 2.** Following the Order of Operations, multiply the 3 and the 5.

$y = 15 + 4$

**Step 3.** Complete the addition to get your answer.

$y = 19$

E.g. 3. Sometimes you will have more than one variable that you will need to substitute.

Substitute $x=6$ and $y=-3$ into the expression $\dfrac{24}{x}(y+5)$ then simplify.

**Step 1.** Replace the variables with the numbers. No $\times$ sign needed this time because there are no coefficients.

$$\frac{24}{6}(-3+5)$$

**Step 2.** Following the Order of Operations, add the −3 and the 5.

$$=\frac{24}{6}(2)$$

**Step 3.** Complete the division and the multiplication to get your answer.

$$=4(2)=4\times 2=8$$

**Please note: In algebra, and mathematics in general, parentheses means to multiply.**

So therefore $4(2)$ is the same as $4\times 2$ which equals 8.

Also note that you must be reasonably confident with negative numbers as there are many expressions with negative terms and/or negative variables and many times you will need to substitute negative numbers into these variables.

**Complete the following substitutions**

a) $2a+3b$ $(a=4, b=5)$

b) $3x-4y$ $(x=2, y=1)$

c) $3b-ab$ $(a=4, b=5)$

d) $\dfrac{16}{y}-7+x$ $(x=5, y=4)$

e) $4(x+y)$ $(x=5, y=4)$

f) $3x+4y$ $(x=-2, y=-3)$

g) $7y-3+4x$ $(x=-3, y=-2)$

h) $4(x-y)$ $(x=-5, y=-4)$

i) $\dfrac{15}{e}-\dfrac{18}{d}$ $(d=-2, e=5)$

j) $5f+\dfrac{8g}{16}$ $(f=-2, g=5)$

You must never forget to follow the Order of Operations, Especially if there are exponents or powers involved in the substitution process.

Remember PEMDAS or BODMAS

**P**arentheses
**E**xponents
**M**ultiplication
**D**ivision
**A**ddition
**S**ubtraction

Or

**B**rackets
**O**rders (Powers, Exponents etc)
**D**ivision
**M**ultiplication
**A**ddition
**S**ubtraction

Examples,

E.g. 1. *Simplify* $9a^3$ when $a=2$

**Step 1.** Replace the variable with the number. Remember to put the × sign in.

$9 \times 2^3$

**Step 2.** Complete the exponent.

$= 9 \times 8$

**Step 3.** Complete the multiplication.

$= 72$

E.g. 2. *Simplify* $5a^2 + \dfrac{b^3}{100} - 5$ when $a = 4$ and $b = -10$

**Step 1.** Replace the variables with the numbers. Remember to put the × sign in.

$5 \times 4^2 + \dfrac{(-10)^3}{100} - 5$

**Step 2.** Complete the exponents.

$= 5 \times 16 + \dfrac{-1000}{100} - 5$

**Step 3.** Complete the multiplication and division before the addition and subtraction.

$= 80 - 10 - 5$

$= 70 - 5$

$= 65$

In the question above, parentheses were placed around the $-10$ because it is important to remember that the $(-)$ sign must stay with the 10.

**Complete the following substitutions with exponents.**
Substitute $x = 2$ and $y = -3$ for each expression, then simplify.

a) $y^2$

b) $5x^2$

c) $12y^2$

d) $2x^3$

e) $2y^3 - y^2$

f) $6x^4 - 2y^2$

g) $\dfrac{x^4}{8}$

h) $\dfrac{y^3}{9}$

i) $\dfrac{10x^3}{80}$

j) $3y^2 + \dfrac{x^5}{2} - 4$

# Chapter 5

**Exponents (Powers) in Algebra**

As mentioned earlier, an exponent or power is the small number written to the top right of a number or variable e.g. $5^2$ or $a^2$. The larger bottom number is called the ***base***. The power tells you how many times the base is multiplied by itself.

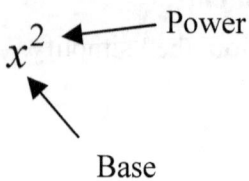

When multiplying a number by itself, the answer can be written as a larger number, or as the same number with an exponent or power.

Examples.

$3 \times 3 = 9$ or $3 \times 3 = 3^2$

$4 \times 4 \times 4 = 64$ or $4 \times 4 \times 4 = 4^3$

But when a variable is multiplied by itself, the answer can only be written with an exponent.

$a \times a = a^2$

$x \times x \times x \times x \times x = x^5$

When a variable or number with an exponent is multiplied out, such as $x \times x \times x \times x \times x$, this is referred to as the **expanded form**.

In algebra there is always a need to work with exponents so it is important to know the rules that apply to them.

The following are the rules you will need to know. These rules are not in any particular order.

**Rule 1.** Any base to the power $0$ is equal to $1$.

E.g. 1. $a^0 = 1$

E.g. 2. $x^0 = 1$

E.g. 3. $8^0 = 1$

E.g. 4. $(x^2 y^3 z^4)^0 = 1$


**Rule 2.** Any base to the power of $1$ doesn't change.

E.g. 1. $a^1 = a$

E.g. 2. $x^1 = x$

E.g. 3. $7^1 = 7$


**Rule 3.** One $(1)$ to any power still equals $1$.

E.g. 1. $1^8 = 1$

E.g. 2. $1^{25} = 1$

E.g. 3. $1^{1000} = 1$

**Rule 4.** Any base with a negative exponent gets written as a **reciprocal** with a positive exponent.

E.g. 1. $a^{-1} = \dfrac{1}{a}$

E.g. 2. $x^{-3} = \dfrac{1}{x^3}$

E.g. 3. $4^{-2} = \dfrac{1}{4^2} = \dfrac{1}{16}$

E.g. 4. $\dfrac{1}{x^{-3}} = x^3$

This is so because, $\dfrac{1}{x^{-3}} = 1 \div x^{-3} = \dfrac{1}{1} \div \dfrac{1}{x^3} = \dfrac{1}{1} \times \dfrac{x^3}{1} = x^3$

**Rule 5.** When multiplying expressions with exponents, that have the same base, the base stays the same and you **add** the exponents.

E.g. 1. $a^4 \times a^3 = a^{4+3} = a^7$

This is because $a^4 = a \times a \times a \times a$ in expanded form and $a^3 = a \times a \times a$ in expanded form. When these are multiplied together, you get $\underline{a^4} \times \underline{a^3} = \underline{a \times a \times a \times a} \times \underline{a \times a \times a} = a^{4+3} = a^7$ (as you can see there are seven $a$ s)

E.g. 2. $x^5 \times x^4 \times x^2 = x^{5+4+2} = x^{11}$

E.g. 3. $2^3 \times 2^2 = 2^{3+2} = 2^5 = 32$

If the base is a number like the example above, you can continue to work it out.

**Rule 6.** When dividing expressions with exponents, that have the same base, the base stays the same and you **subtract** the exponents.

E.g. 1. $a^7 \div a^4 = a^{7-4} = a^3$

Remember that a division in algebra should be rewritten as a fraction. If the above calculation is written in expanded form, you get:

$$\frac{a^7}{a^4} = \frac{a \times a \times a \times a \times a \times a \times a}{a \times a \times a \times a}$$

This way you can see that you can cancel away the same number of variables from the numerator and the denominator. In this case, you cancel away 4 from the top and 4 from the bottom, (because there are only 4 on the bottom).

$$= \frac{a \times a \times a \times \cancel{a} \times \cancel{a} \times \cancel{a} \times \cancel{a}}{\cancel{a} \times \cancel{a} \times \cancel{a} \times \cancel{a}}$$

When canceling, you ask yourself "how many times does $a$ go into $a$" the answer is 1, therefore you can replace the $a$s you cross out with 1s.

$$= \frac{a \times a \times a \times 1 \times 1 \times 1 \times 1}{1 \times 1 \times 1 \times 1}$$

$$= \frac{a \times a \times a}{1}$$

Since you don't need all the 1s, you don't need to write them in the final answer.

$$=a^3$$

This was the long version to explain what happens when you simplify. It does look confusing, but you won't need to do this. You will follow the method below.

To avoid writing out in expanded form every time, you would just use subtraction.

$$\frac{a^7}{a^4} = \frac{a^{\cancel{7}3}}{\cancel{a^4}} = \frac{a^3}{1} = a^3$$

Note that when you subtract the 4 from the 7, you've used up all the $a$s in the denominator (like in the long version above) and this gets simplified to a 1 (and you don't need to write the 1 in the final answer).

E.g. 2. $x^5 \div x^2 = x^{5-2} = x^3$

E.g. 3. $\dfrac{22c^3}{11c^7} = \dfrac{22 \times c^3}{11 \times c^7} = \dfrac{2}{c^4}$

**Rule 7.** When raising a power to another power, you keep the base the same and multiply the powers.

E.g. 1. $(a^3)^2 = a^{3 \times 2} = a^6$

This happens because of **Rule 5** above.

$(a^3)^2$ is the same as $a^3 \times a^3$ which of course equals $a^6$.

E.g. 2. $(x^2 y^3)^4 = x^{2 \times 4} y^{3 \times 4} = x^8 y^{12}$

E.g. 3. $(2^2)^3 = 2^{2 \times 3} = 2^6 = 64$

**Rule 8.** When there is a single power outside parentheses, everything inside the parentheses is raised to that power.

E.g. 1. $(abc)^4 = a^4 b^4 c^4$

E.g. 2. $(rst)^5 = r^5 s^5 t^5$

This applies for fractions as well.

E.g. 3. $(\dfrac{x}{y})^3 = \dfrac{x^3}{y^3}$

**Rule 9.** If the exponent or power is a fraction, then you must find the root of the base, depending on the value of the fraction.

E.g. 1. $a^{\frac{1}{2}} = \sqrt{a}$

E.g. 2. $x^{\frac{1}{3}} = \sqrt[3]{x}$

E.g. 3. $x^{\frac{2}{3}} = (\sqrt[3]{x})^2$

# Chapter 6

## Multiplying Algebraic Terms

When numbers are multiplied, larger numbers are made, for instance 2 × 3 = 6. However when variables are multiplied, they are just put next to each other. So $a \times b = ab$ and $x \times y = xy$ and $r \times s \times t = rst$ etc.

If the variables that are multiplied are the same, as in $x \times x$, then exponents or powers are used to show this. In this case $x \times x = x^2$.

The steps you take to multiply in algebra are:

**Step 1.** Multiply the coefficients

**Step 2.** Multiply the variables

**Step 3.** Put the two together.


Examples

E.g. 1. For a question like multiply $4x \times 5y$,

**Step 1.** Multiply the coefficients

$4 \times 5 = 20$

**Step 2.** Multiply the variables

$x \times y = xy$

**Step 3.** Put the two together.

$20xy$

E.g. 2. If your question has a few terms like $2a \times 5b \times c \times 4e$,

**Step 1.** Multiply the coefficients

$2 \times 5 \times 1 \times 4 = 40$ (remember the coefficient of $c$ is 1)

**Step 2.** Multiply the variables

$a \times b \times c \times e = abce$

**Step 3.** Put the two together.

$40abce$

E.g. 3. If every term doesn't have a variable like $3 \times 5d$,

**Step 1.** Multiply the coefficients

$3 \times 5 = 15$

**Step 2.** Multiply the variables

There is only a $d$ in this case.

**Step 3.** Put the two together.

$15d$

E.g. 4. If the variables are the same like $8y \times 3y$,

**Step 1.** Multiply the coefficients

$8 \times 3 = 24$

**Step 2.** Multiply the variables

$y \times y = y^2$

**Step 3.** Put the two together.

$24y^2$

**Multiply these algebraic terms:**

a) $8 \times 3y$

b) $9x \times 3y$

c) $-7a \times 2b$

d) $2u \times 5$

e) $11v \times 4t$

f) $9s \times 4t^2$

g) $6a \times 3a$

h) $10x \times 3xy$

i) $4xy \times 3y \times 2x$

j) $10d \times 4 \times 2ef$

# Chapter 7

## Dividing Algebraic Terms

As you already know, fractions are divisions, and therefore divisions are fractions.

In algebra, whenever you see a division, you should immediately rewrite it as a fraction.

For instance, you know that $9 \div 10$ is the same as $\frac{9}{10}$. In algebra, you would do the same thing. If you had an expression like $9a \div 10b$ you would immediately rewrite this as $\frac{9a}{10b}$, and $x \div y$ would be written as $\frac{x}{y}$.

**Rewrite these divisions as fractions.**

a) $5c \div 8d$

b) $x \div -5x$

c) $25b \div 5$

d) $16c \div 4c$

e) $-30c \div 6$

f) $32xy \div 16$

g) $45c \div -15c$

h) $28fg \div 7g$

i) $abc \div def$

j) $ac \div -8c$

After rewriting the division as a fraction, the next thing you need to do is to simplify it. The easiest way to do this is to simplify the coefficients (numbers), then the variables (letters). This is done separately to avoid confusion. As you get used to this, you will be able to simplify more quickly.

Simplifying the numbers is the same as in fractions, look for the common factor and simplify. To simplify the variables you would cancel (by crossing out) whatever is in the numerator with whatever is the same in the denominator. This is also a common factor.

Examples.

E.g. 1. Simplify $28fg \div 7g$.

**Step 1.** Rewrite as a fraction and separate the numbers and letters with a (×) sign.

$$= \frac{28 \times fg}{7 \times g}$$

**Step 2.** Simplify the numbers by finding common factors and cross out whatever is the same for the letters.

$$= \frac{\overset{4}{\cancel{28}} \times f \cancel{g}}{\cancel{7} \times \cancel{g}}$$

**Step 3.** Rewrite the simplified term with the new values.

$$= 4f$$

E.g. 2. Simplify $5a^4b^6 \div 6ab^2$.

**Step 1.** Rewrite as a fraction and separate the numbers and letters with a (×) sign.

$$= \frac{5 \times a^4 b^6}{6 \times ab^2}$$

**Step 2.** Simplify the numbers by finding common factors and cross out whatever is the same for the letters.

$$= \frac{5 \times a^{\cancel{4}3} b^{\cancel{6}4}}{6 \times \cancel{a}\cancel{b^2}}$$

Notice that the numbers do not change because there are no common factors. The exponents however can be simplified by using **Rule 6** above.

**Step 3.** Rewrite the simplified term with the new values.

$$= \frac{5a^3 b^4}{6}$$

**Simplify the following**

a) $a^8 \div a^3$

b) $9x^4 \div x^2$

c) $\dfrac{14x^7}{7x^2}$

d) $\dfrac{18x^3}{6x^7}$

e) $\dfrac{11x^9}{33x^3}$

f) $\dfrac{14x^6 y^8}{10x^5 y^5}$

g) $-45r^4 s^4 \div 15r^3 s^4$

h) $81a^{12} b^8 c \div -36a^{15} b^9 c^2$

i) $\dfrac{-g^9 h^8}{8g^9 h^8}$

j) $\dfrac{36x^{11} z^4}{27x^7 y^3 z^9}$

# Chapter 8

**Multiplying and Simplifying Algebraic Expressions**

Or

**Expanding and Simplifying Brackets**

Or

**The Distributive Law (or Property)**

You know from the Order of Operations rule that you work on Parentheses or Brackets first (PEMDAS) or (BODMAS).

The Distributive Law allows you to jump straight to the multiplication, but you must follow certain rules.

When you are using the Distributive Law, you are taking the terms out of brackets or parentheses.

## Expanding Expressions of the Type $a(b+c)$ (two factors)

Look at this example.

*Solve* $7(4 + 5)$

Following the Order of Operations rule, you complete the work inside the parentheses first $4 + 5 = 9$, then you multiply with the number outside the parentheses $7(9) = 63$.

If you were told to use the distributive law to answer this question, then you must multiply the number outside the parentheses with each number inside the parentheses separately, then add.

Drawing in the lines above the numbers like $\overline{7(4+5)}$ shows you what needs to be multiplied together (after a while you won't need to draw in the lines, you'll be able to do it in your head).

7 × 4 + 7 × 5 (parentheses are removed)
= 28 + 35
= 63

The answer is the same as following the first method.

In algebra, when you have an expression like $4(x+3)$, you cannot follow the Order of Operations because you don't know what the value of $x$ is. Therefore you would use the **distributive law**.

Examples

E.g. 1. Expand and simplify $4(x+3)$

**Step 1.** Place the lines over the expression to show what you are multiplying.

$\overline{4(x+3)}$

**Step 2.** Multiply out the terms, following the lines and remove the parentheses.

$= 4 \times x + 4 \times 3$

**Step 3.** Collect any like terms and simplify.

$= 4x + 12$

E.g. 2. Expand and simplify $3a(a+5b)$

**Step 1.** Place the lines over the expression to show what you are multiplying.

$\overparen{3a(a+5b)}$

**Step 2.** Multiply out the terms, following the lines and remove the parentheses.

$=3a\times a + 3a\times 5b$

**Step 3.** Collect any like terms and simplify.

$=3a^2+15ab$

E.g. 3. Expand and simplify $5(a+3b)+a(8-9a)-6b$

**Step 1.** Place the lines over the expression to show what you are multiplying.

$\overparen{5(a+3b)}+\overparen{a(8-9a)}-6b$

**Step 2.** Multiply out the terms, following the lines and remove the parentheses.

$5a+15b+8a-9a^2-6b$

**Step 3.** Collect any like terms and simplify.

$13a+9b-9a^2$

**Expand and simplify the following**

a) $4(x+10)$

b) $3b(c-d)$

c) $2(3x+4)$

d) $5x(x+2y)$

e) $6a^2(-2ab+a)$

f) $8x^3y(x^4+3x^2y^3)$

g) $5d(2-e)+8de$

h) $7q(3-8r)-28q$

i) $2a^2(4a^3-4)+a(8a^5-1)$

j) $3x^2(5x^3-2)-4x^4(9x-3x^3)$

## Expanding Expressions of the Type $(a+b)(c+d)$, (two factors)

Look at these examples

E.g. 1. Expand and simplify $(x+2)(x+3)$

As there are two sets of parentheses, each term inside the first set of parentheses is multiplied with each term inside the second set of parentheses. Drawing in the lines above the numbers, like you did in the previous section, shows you what needs to be multiplied together, but this time you will need two sets of lines, one on top and one on the bottom like so:

**Step 1.** Place the lines over the expression to show what you are multiplying.

$(x+2)(x+3)$

**Step 2.** Multiply out the terms, following the lines and remove the parentheses.

$= x \times x + x \times 3 + 2 \times x + 2 \times 3$

$= x^2 + 3x + 2x + 6$

**Step 3.** Collect like terms and simplify.

$= x^2 + 5x + 6$

Note that whenever you need to expand two parentheses, you can follow this method.

E.g. 2. Expand and simplify $(2x+5)(3x+2)$

**Step 1.** Place the lines over the expression to show what you are multiplying.

$(2x+5)(3x+2)$

**Step 2.** Multiply out the terms, following the lines and remove the parentheses.

$= 2x \times 3x + 2x \times 2 + 5 \times 3x + 5 \times 2$

$= 6x^2 + 4x + 15x + 10$

**Step 3.** Collect like terms and simplify.

$= 6x^2 + 19x + 10$

**Expand and simplify these $(a+b)(c+d)$ type expressions**

a) $(x+1)(x+3)$

b) $(x+4)(x-5)$

c) $(x-8)(x-6)$

d) $(2x+3)(x-5)$

e) $(x+2)(3x-6)$

f) $(3x+1)(x+2)$

g) $(4x+2)(x-3)$

h) $(3x+5)(2x-7)$

i) $(2x+9)(4x+7)$

j) $(6x-5)(2-3x)$

## Expanding Perfect Squares $(a+b)^2$ type Expressions

Sometimes you will be asked to expand a **Perfect Square.**

A perfect square is a number, or variable or term or expression that is multiplied by itself. The following are perfect squares:

$5^2$, $b^2$, $(3x)^2$, $(2x+1)^2$.

E.g. 3a. Expand and simplify $(2x+1)^2$

**Step 1.** Write out the expression in full without the exponent

$(2x+1)(2x+1)$

**Step 2.** Place the lines over the expression to show what you are multiplying.

$(2x+1)(2x+1)$

**Step 3.** Multiply out the terms, following the lines and remove the parentheses.

$= 2x \times 2x + 2x \times 1 + 1 \times 2x + 1 \times 1$

$= 4x^2 + 2x + 2x + 1$

**Step 4.** Collect like terms and simplify.

$= 4x^2 + 4x + 1$

There is a shortcut to expanding perfect squares, but you should probably use the method above until you can answer these type of questions easily.

For the shortcut, the same expression as above will be used to show the difference.

E.g. 3b. Expand and simplify $(2x+1)^2$

**Step 1.** Square the first term in the parentheses

$(2x)^2 = 4x^2$

**Step 2.** Multiply the first and last terms with each other then multiply this answer by 2

$2x \times 1 \times 2 = 4x$

**Step 3.** Square the last term in the parentheses

$1^2 = 1$

**Step 4.** Put all three answers together (take note of any positive or negative signs)

$4x^2 + 4x + 1$

Your answer is the same as in E.g. 3a above.

E.g. 3c. Expand and simplify $(3x-2)^2$ (note the $-2$).

**Step 1.** Write out the expression in full without the exponent

$(3x-2)(3x-2)$

**Step 2.** Place the lines over the expression to show what you are multiplying.

$(3x-2)(3x-2)$

**Step 3.** Multiply out the terms, following the lines and remove the parentheses. (Remember the rules for Negative Numbers.)

$$= 3x \times 3x + 3x \times -2 - 2 \times 3x - 2 \times -2$$

$$= 9x^2 - 6x - 6x + 4$$

**Step 4.** Collect like terms and simplify.

$$= 9x^2 - 12x + 4$$

When expanding perfect squares:

The final answer will have two plus (+) signs if the perfect square has a plus (+) sign.

The final answer will have a minus (−) sign and a plus (+) sign if the perfect square has a minus (−) sign.

Expand and Simplify these Perfect Squares $(a+b)^2$ type.

a) $(x+1)^2$

b) $(x-3)^2$

c) $(x+7)^2$

d) $(2x-2)^2$

e) $(3x+4)^2$

f) $(2x-4)^2$

g) $(3x-5)^2$

h) $(2-3x)^2$

i) $(2x+6)(2x+6)$

j) $(6y-5x)(6y-5x)$

# Expanding Difference Of Two Squares $(a+b)(a-b)$, (two factors)

Notice the terms inside the parentheses are the same as each other except for the operation signs. One is a (+) and the other is a (−). This expression is called a Difference of Two Squares because when it is expanded, the answer will have two perfect squares separated by a minus (−) sign.

E.g. 4a. Expand and simplify $(3x+2)(3x-2)$

**Step 1.** Place the lines over the expression to show what you are multiplying.

$$(3x+2)(3x-2)$$

**Step 2.** Multiply out the terms, following the lines and remove the parentheses.

$= 3x \times 3x \ + 3x \times -2 \ + 2 \times 3x \ + 2 \times -2$

$= 9x^2 - 6x + 6x - 4$

**Step 3.** Collect like terms and simplify.

$= 9x^2 - 4$

Note that there is no middle term in the final answer. This is because $-6x+6x$ cancel each other out and there is no need to put a 0 in the expression.

This method also has a shortcut. Make sure you know the longer version well before you use this method.

**E.g. 4b.** Using the same expression $(3x+2)(3x-2)$

**Step 1.** Square the first term

$(3x)^2 = 9x^2$

**Step 2.** Square the last term

$2^2 = 4$

**Step 3.** Put the two together with a minus (−) sign between them

$9x^2 - 4$

You see that the answer is the same with less working out.

**Expand and Simplify these D.O.T.S. $(a+b)(a-b)$ type Expressions**

a) $(x+3)(x-3)$

b) $(x+9)(x-9)$

c) $(7-x)(7+x)$

d) $(3-6x)(3+6x)$

e) $(5x+4)(5x-4)$

f) $(8-7x)(8+7x)$

g) $(4-x)(x+4)$

h) $(8-2a)(2a+8)$

i) $(4x+2y)(4x-2y)$

j) $(2y+7x)(2y-7x)$

# Expanding Three Factors

When expanding expressions with three factors, you would expand two factors first, then you would multiply with the third factor.

Examples

E.g. 5a. Expand and simplify $x(x+2)(x-3)$

**Step 1.** Expand the parentheses first but leave your answer in parentheses. You do this to avoid confusion with the $x$ outside the parentheses.

$x(x^2 - x - 6)$

**Step 2.** Place the lines over the expression to show that you are multiplying the $x$ outside the parentheses with the terms inside the parentheses.

$x(x^2 - x - 6)$

**Step 3.** Multiply out the terms following the lines and remove the parentheses.

$x^3 - x^2 - 6x$

This is the final answer as there are no like terms.

If you have three sets of parentheses, the process is similar.

E.g. 5b. Expand and simplify $(x+1)(x+2)(x+3)$

**Step 1.** Expand the second and third parentheses first but leave your answer in parentheses.

$(x+1)(x^2 + 5x + 6)$

**Step 2.** Place the lines over the expression to show what you are multiplying.

$$(x+1)(x^2+5x+6)$$

This time there will be six terms

**Step 3.** Multiply out the terms following the lines and remove the parentheses.

$x \times x^2 \ + x \times 5x \ + x \times 6 \ + 1 \times x^2 \ + 1 \times 5x \ + 1 \times 6$

$= x^3 + 5x^2 + 6x + x^2 + 5x + 6$

**Step 4.** Collect like terms and simplify to get your final answer

$= x^3 + 6x^2 + 11x + 6$

This same process can be followed to expand expressions with more parentheses.

**Expand and Simplify the Following**

a) $x(x+4)(x+5)$

b) $x(x-1)(x-3)$

c) $-x(x+3)(x-4)$

d) $2x(x+6)(x-8)$

e) $-3x(x-2)(x-1)$

f) $(x+4)(x+2)(x+5)$

g) $(x+2)(x-5)(x-1)$

h) $(x-3)(x-2)(x-6)$

i) $(2x+1)(x-4)(x+3)$

j) $(x+4)(2x-2)(x+3)$

# Chapter 9

## Factoring (Factorising) Algebraic Expressions

Factoring, or factorising is the reverse process of expanding.

When factoring a binomial expression, you are looking for the greatest common factor (GCF) or highest common factor (HCF) to divide into the two terms. The GCF is taken out and the remainder is put into parentheses.

Knowing your multiplication tables will help you to find the GCF of the coefficients (numbers) of each term. Knowing your exponent rules will help you find the GCF of the variables in each term. Unfortunately there is no short cut. You just have to learn them and recognize them.

**Binomial Expressions**

Examples

E.g. 1. Factor the expression $10a-12$.

**Step 1.** Find the GCF of the two term of the expression

The GCF is 2.

**Step 2.** Divide each term by the GCF. Use the exponent rules when needed (be careful of the $+$ and $-$ signs).

$$\frac{\cancel{10}^5 a}{\cancel{2}_1} - \frac{\cancel{12}^6}{\cancel{2}_1} = 5a - 6$$

**Step 3.** Place the GCF outside parentheses and the answer to step 2 inside the parentheses for the final answer.

$2(5a-6)$

E.g. 2. Factor the expression $3xy+3x$

**Step 1.** Find the GCF of the two term of the expression

The GCF is $3x$.

**Step 2.** Divide each term by the GCF. Use the exponent rules when needed (watch the $+$ and $-$ signs).

$$\frac{\cancel{3x} \times y}{\cancel{3x}} + \frac{\cancel{3x}^{1}}{\cancel{3x}_{1}} = y+1$$

**Step 3.** Place the GCF outside parentheses and the answers to step 2 inside the parentheses for the final answer.

$3x(y+1)$

E.g. 3. Factor the expression $x^3+x^7$

**Step 1.** Find the GCF of the two term of the expression

The GCF is $x^3$.

**Step 2.** Divide each term by the GCF. Use the exponent rules when needed (watch the $+$ and $-$ signs).

$$\frac{\cancel{x^3}}{\cancel{x^3}} + \frac{x^{\cancel{7}4}}{\cancel{x^3}} = 1+x^4$$

**Step 3.** Place the GCF outside parentheses and the answers to step 2 inside the parentheses for the final answer.

$x^3(1+x^4)$

E.g. 4. Factor the expression $10x^2y^3 - 15xy^4$

**Step 1.** Find the GCF of the two term of the expression

The GCF is $5xy^3$.

**Step 2.** Divide each term by the GCF. Use the exponent rules when needed (watch the $+$ and $-$ signs).

$$\frac{\cancel{10}^2 \times \cancel{x^2} \times \cancel{y^3}}{\cancel{5} \times \cancel{x} \times \cancel{y^3}} - \frac{\cancel{15}^3 \times \cancel{x} \times \cancel{y^4}}{\cancel{5}_1 \times \cancel{x} \times \cancel{y^3}} = 2x - 3y$$

**Step 3.** Place the GCF outside parentheses and the answers to step 2 inside the parentheses for the final answer.

$5xy^3(2x-3y)$

As you can see from all the crossing out, this work can get messy and confusing, but it is important that you persist with this until you get used to the process. Just make sure your work is neat.

**Factor the following binomial expressions**

a) $3x - 45$ 

c) $4xy - 18$

b) $20 - 4x$ 

d) $x^2 - 4x$

e) $12xy - 4x^2$

f) $7ax^2 - 2a^2x$

g) $4ab^2 - 4ab$

h) $15xy + 30x$

i) $40xy - 30x$

j) $32xy + 36y$

# Factoring using the Difference of Two Squares (D.O.T.S.) method

The Difference of Two Squares rule says that $a^2-b^2=(a+b)(a-b)$ and, you know that when an expression like $(x+3)(x-3)$ is expanded, the result is $x^2-9$. Therefore to factor the expression $x^2-9$, you would put the square root of each term in two sets of parentheses, one with a plus (+) sign and the other with a minus (−) sign.

To be successful at these types of questions, it helps to know your **perfect squares**.

Examples

E.g. 1. Factor $x^2-9$

**Step 1.** Look for common factors. If there are any, put them outside of parentheses and the remaining terms inside

$x^2-9$ has no common factors

**Step 2.** Write each term as a square

$=(x)^2-(3)^2$

**Step 3.** Place each term from Step 2 above in two sets of parentheses one with a plus (+) and one with a minus (−).

$=(x+3)(x-3)$

E.g. 2. Factor $16k^2 - 25b^2$

**Step 1.** Look for common factors. If there are any, put them outside of parentheses and the remaining terms inside

$16k^2 - 25b^2$ has no common factors

**Step 2.** Write each term as a square

$= (4k)^2 - (5b)^2$

**Step 3.** Place each term from Step 2 above in two sets of parentheses one with a plus (+) and one with a minus (−).

$= (4k + 5b)(4k - 5b)$

E.g. 3. Factor $50a^2 - 72m^2$

**Step 1.** Look for common factors. If there are any put them outside of parentheses and the remaining terms inside

$50a^2 - 72m^2$ has the GCF of 2

$= 2(25a^2 - 36m^2)$

**Step 2.** Write each term as a square. You must remember to include the common factor.

$= 2[(5a)^2 - (6m)^2]$

**Step 3.** Place each term from Step 2 above in two sets of parentheses one with a plus (+) and one with a minus (−).

$= 2(5a + 6m)(5a - 6m)$

E.g. 4. Factor $(x+3)^2 - 16$

**Step 1.** Look for common factors. If there are any put them outside of parentheses and the remaining terms inside

$(x+3)^2 - 16$ has no common factors

**Step 2.** Write each term as a square

$= (x+3)^2 - (4)^2$ note that the first term is already written as a square.

**Step 3.** Place each term from Step 2 above in two sets of parentheses one with a plus (+) and one with a minus (−).

$= (x+3+4)(x+3-4)$

**Step 4.** Collect like terms for the final answer

$= (x+7)(x-1)$


E.g. 5. Factor $x^2 - 7$

**Step 1.** Look for common factors. If there are any put them outside of parentheses and the remaining terms inside
$x^2 - 7$ has no common factors

**Step 2.** Write each term as a square

$(x)^2 - (\sqrt{7})^2$

**Step 3.** Place each term from Step 2 above in two sets of parentheses one with a plus (+) and one with a minus (−).
$(x+\sqrt{7})(x-\sqrt{7})$

Sometimes you just leave your answer as a **surd**. This means that if the number isn't a perfect square, as in the 7 above, then you just leave it like that with a square root sign.

E.g. 4. Factor $(x+5)^2-(x+3)^2$

**Step 1.** Look for common factors. If there are any put them outside of parentheses and the remaining terms inside

$(x+5)^2-(x+3)^2$ has no common factors

**Step 2.** Write each term as a square

$=(x+5)^2-(x+3)^2$ note that both terms are already written as squares

**Step 3.** Place each term from Step 2 above in two sets of parentheses one with a plus (+) and one with a minus (−).

$=(x+5+x+3)(x+5-x-3)$ note the change in sign from positive to negative

**Step 4.** Collect like terms

$=(2x+8)(2)$

Step 5. If you can factor the expression further, you should do so.

$=2(2x+8)$ the common factor of 2 can be taken out of the parentheses. This is multiplied with the term outside the parentheses.

$=2\times 2(x+4)$

$=4(x+4)$

**Factor the following using the D.O.T.S. method**

a) $b^2 - 4$

b) $49x^2 - y^2z^2$

c) $7a^2 - 7b^2$

d) $3a^2b^2 - 12c^2$

e) $(x+3)^2 - 4$

f) $1 - (3x-2)^2$

g) $3(x-4)^2 - 12$

h) $x^2 - 3$

i) $27a^2 - 15$

j) $(5x-4)^2 - (x-7)^2$

## Factoring using the Perfect Square method

The Perfect Square rule says:

$a^2 + 2ab + b^2 = (a+b)^2$ and

$a^2 - 2ab + b^2 = (a-b)^2$

Notice that the second rule has negatives.

To factor using the Perfect Square rule, you would reverse the process that was used for expanding. You must first determine if the trinomial can be factored with this method. Look at the first and last terms and if they are perfect squares, then it may be possible. Don't forget you may need to take out a common factor first.

**<u>Be careful with negatives.</u>**
**<u>If the trinomial has two plus (+) signs, then the answer will have a plus (+) sign (see the rule above).</u>**
**<u>If the trinomial has a minus (−) sign and a plus (+) sign, then the answer will have a minus (−) sign (see the rule above).</u>**

E.g. 1. Factor $x^2 + 10x + 25$

**Step 1.** Look for any common factors. Take note of the operation signs.

$x^2 + 10x + 25$ has no common factors

**Step 2.** Determine if the first and last terms are perfect squares and if so write them as squares

$(x)^2 + 10x + (5)^2$

**Step 3.** Decide if the middle term is 2 times the first term times the last term from the previous step.

$2 \times x \times 5 = 10x$ this is the middle term

**Step 4.** Since the trinomial has two (+) signs then your answer will have a (+) sign

$(x+5)^2$

Note the two terms in the answer are the same as the squares from Step 2.

E.g. 2. Factor $4x^2 - 12x + 9$

**Step 1.** Look for any common factors. Take note of the operation signs.

$4x^2 - 12x + 9$ has no common factors

**Step 2.** Determine if the first and last terms are perfect squares and if so write them as squares

$(2x)^2 - 12x + (3)^2$

**Step 3.** Decide if the middle term is 2 times the first term times the last term from the previous step.

$2 \times 2x \times 3 = 12x$ is the middle term.

**Step 4.** Since the trinomial has a (−) sign and a (+) signs then your answer will have a (−) sign

$(2x-3)^2$

Note the two terms in the answer are the same as the squares from Step 2.

E.g. 3 Factor $3x^2-18x+27$

**Step 1.** Look for any common factors. Take note of the operation signs.

$3x^2-18x+27$ has the common factor 3 which needs to be taken out.

$3(x^2-6x+9)$

**Step 2.** Determine if the first and last terms inside the parentheses are perfect squares and if so write them as squares

$3[(x)^2-6x+(3)^2]$

**Step 3.** Decide if the middle term is 2 times the first term times the last term from the previous step.

$2\times x\times 3=6x$ this is the middle term

**Step 4.** Since the trinomial has a (−) sign and a (+) signs then your answer will have a (−) sign

$3(x-3)^2$ make sure you include the common factor.

Note the two terms in the answer are the same as the squares from Step 2.

**Factor the following using the Perfect Square method**

a) $x^2+6x+9$

b) $x^2+4x+4$

c) $x^2-14x+49$

d) $x^2-8x+16$

e) $x^2+2x+1$

f) $4x^2+20x+25$

g) $9x^2+24x+16$

h) $25x^2-10x+1$

i) $3x^2+36x+108$

j) $50x^2-36x+32$

## Factoring by Grouping

Some expressions have four terms. These can sometimes be factored by grouping them. The groups are either **Two and Two**, or **Three and One**. To factor these, you will need to use:

the common factor method,

and/or

the D.O.T.S. method,

and /or

the Perfect Square method.

## Factoring by Grouping Two and Two

To factor by grouping two and two, you need to look at the expression and look for:

i) common factors

ii) D.O.T.S.

Examples

E.g. 1a. Factor $x^2 - 4x + 5x - 20$

**Step 1.** Put a line down the middle of the expression to separate into 2 groups,

$x^2 - 4x \mid +5x - 20$

**Step 2.** Look for a common factor and/or D.O.T.S.,

$= x(x-4)+5(x-4)$ both groups have common factors.
One set of parentheses for each group **must** be the same. If they are not you've made a mistake somewhere.

**Step 3.** Take out the new common factor for both groups and put the remainder in parentheses.

$=(x-4)(x+5)$ The new common factor in this case is $(x-4)$ as it appears in both groups, and the remainder is $(x+5)$.

E.g. 2. Factor $x^2 - y^2 + 5x - 5y$

**Step 1.** Put a line down the middle of the expression to separate into 2 groups,

$x^2 - y^2 \mid + 5x - 5y$

**Step 2.** Look for a common factor and/or D.O.T.S.,

$(x+y)(x-y)+5(x-y)$ the left hand group is factored using D.O.T.S., the right hand group has a common factor

One set of parentheses for each group **must** be the same. If they are not you've made a mistake somewhere.

**Step 3.** Take out the new common factor for both groups and put the remainder in parentheses.

$=(x-y)(x+y+5)$ The new common factor in this case is $(x-y)$ and the remainder is $(x+y+5)$.

# Factoring by Grouping Three and One

To factor by grouping three and one, you need to look at the expression and look for

i) three terms that would satisfy the Perfect Square rule,
ii) the fourth term is a square, and
ii) D.O.T.S. to finalise the factoring

E.g. 1 Factor $x^2+4x+4-y^2$

**Step 1.** Look to see if three terms satisfy the perfect square rule and the fourth term is a square. If yes, put a line to separate these into groups of three and one

$x^2+4x+4 | -y^2$ this expression satisfies the three and one grouping

**Step 2.** Factor the group of three using the perfect square method

$(x+2)^2 - y^2$, now you have a D.O.T.S.

**Step 3.** Factor using the D.O.T.S. method

$(x+2+y)(x+2-y)$

**Factor the following by grouping Two and Two**

a) $ab+3b+5a+15$

b) $3x+3y+xz+yz$

c) $8vj-6j-12v+9$

d) $b^2+2b+4b+8$

e) $c^2-4c-3c+12$

f) $12d^2+9d+8d+6$

g) $a^2-c^2+7a-7c$

h) $16b^2-25e^2+10e+8b$

i) $81-49m^2-27n+21mn$

j) $16-9r^2-16s+12rs$

**Factor the following by grouping Three and One**

a) $x^2+8x+16-y^2$

b) $h^2-4h+4-k^2$

c) $a^2+8a+16-b^2$

d) $r^2-10r+25-s^2$

e) $b^2+2b+1-9n^2$

f) $j^2+6j+9-16t^2$

g) $x^2-6xy+9y^2-16$

h) $a^2-10ab+25b^2-36$

i) $4m^2-12m+9-16n^2$

j) $9v^2+30v+25-4j^2$

# Factoring Quadratic Trinomials

A quadratic expression is an expression that has at least one variable raised to the power of 2 as its highest exponent. Examples of quadratic expressions are $a^2$, $x^2-3$, $b^2+2b$, $x^2-3x+2$.

A quadratic trinomial, however, is an expression with three terms, at least one of which has an exponent of 2 as the highest power. So from the list above, only $x^2-3x+2$ is a quadratic trinomial.

In general, a quadratic trinomial has the form $ax^2+bx+c$, where $a$, $b$ and $c$ do not equal zero. If you need to, make sure you rewrite the expression so the term with the power $(ax^2)$ is the first term and the constant $(c)$ is the third or last term.

Factoring quadratic trinomials becomes very easy when you know your multiplication tables and your negative numbers. A copy of the multiplication tables have been included toward the end of this book and you can obtain a copy of the Easy Steps Math Negative Numbers book from Amazon.com.

Examples

E.g. 1. Factor $x^2-11x+30$

**Step 1.** Look for any common factors.

$x^2-11x+30$ has no common factors

**Step 2.** Multiply the coefficient of the first term with the last term

$1 \times 30 = 30$

**Step 3.** Find two numbers that multiply into this number and add into the coefficient of the middle term. List the positive and/or negative factors to help you.

Factors are:

1 and 30
−1 and −30
2 and 15
−2 and −15
3 and 10
−3 and −10
5 and 6
−5 and −6

Choose the factors from the list that

× into 30
+ into −11

These are −5 and −6

Test these before using $-5-6=-11$ and $-5\times-6=30$

**Step 4.** Rewrite the original expression and replace the middle term with these two numbers (don't forget the variable)

$x^2 - 5x - 6x + 30$

**Step 5.** Factor by grouping two and two

$x^2 - 5x \big| -6x + 30$

$= x(x-5) - 6(x-5)$

$=(x-5)(x-6)$ is your final answer.

E.g. 2. Factor $3a^2-11a-4$

**Step 1.** Look for any common factors.

$3a^2-11a-4$ has no common factors

**Step 2.** Multiply the coefficient of the first term with the last term

$3 \times -4 = -12$

**Step 3.** Find two numbers that multiply into this number and add into the coefficient of the middle term. List the positive and/or negative factors to help you.

Factors are:

1 and −12
−1 and 12
2 and −6
−2 and 6
3 and −4
−3 and 4

Choose the factors from the list that

× into −12
+ into −11

These are 1 and −12

Test these before using $+1-12=-11$ and $+1 \times -12 = -12$

**Step 4.** Rewrite the original expression and replace the middle term with these two numbers (don't forget the variable)

$3x^2 + x - 12x + 4$

**Step 5.** Factor by grouping two and two

$3x^2 + x \mid -12x + 4$

$= x(3x+1) - 4(3x+1)$

$= (3x+1)(x-4)$ is your final answer.

E.g. 3. Factor $5d^2 - 25d + 20$

**Step 1.** Look for any common factors.

$5d^2 - 25d + 20$ has the common factor of 5

$5(d^2 - 5d + 4)$

**Step 2.** Multiply the coefficient of the first term with the last term

$1 \times 4 = 4$

**Step 3.** Find two numbers that multiply into this number and add into the coefficient of the middle term. List the positive and/or negative factors to help you.

Factors are:

1 and 4
−1 and −4
2 and 2

–2 and –2

Choose the factors from the list that

× into 4
+ into –5

These are –1 and –4

Test these $-1-4=-5$ and $-1\times-4=4$

**Step 4.** Rewrite the original expression and replace the middle term with these two numbers (don't forget the variable)

$5(d^2-d-5d+4)$

**Step 5.** Factor by grouping two and two

$5(d^2-d|-4d+4)$

$=5[d(d-1)-4(d-1)]$

$=5(d-1)(d-4)$ is your final answer.

**Factor these quadratic trinomials**

a) $x^2+11x+24$

b) $a^2+a-30$

c) $z^2+7z-60$

d) $2a^2+5a+2$

e) $3m^2+8m-16$

f) $2-10j+12j^2$

g) $4b^2+8b+3$

h) $12s^2-11s-15$

i) $3x^2+18x+24$

j) $-5a^2-20a+105$

# Factoring by Completing the Square Method

The completing the square method is used when you cannot factor a trinomial using the quadratic trinomial method. Completing the square requires you to slightly modify the trinomial so that you use both the Perfect Square method and the Difference of Two Squares method, as you did in the grouping by three and one.

Examples

E.g. 1. Factor $x^2+6x-2$

**Step 1.** Take out the common factor if there is one.

$x^2+6x-2$, has no common factors

**Step 2.** Check to see if the factoring quadratic trinomial method works.

$x^2+6x-2$, cannot be factored with the quadratic trinomial method.

**Step 3.** You must create a perfect square. You do this by:

i) halving the coefficient of the middle term $6\div2=3$

ii) then square this number $3^2=9$

iii) then add **and** subtract this number to the expression as the third and fourth terms. You add and subtract the same number so that you don't change the expression. However the first three terms creates a perfect square, which is what you need to continue.

$x^2+6x+9-9-2$

**Step 4.** Put parentheses around the perfect square and work out the last two digits.

$= (x^2+6x+9)-9-2$

$= (x^2+6x+9)-11$

**Step 5.** Factor the Perfect Square, and you will have a D.O.T.S.

$= (x+3)^2 - 11$

$= (x+3)^2 - \sqrt{(11)^2}$

Note that $\sqrt{(11)^2} = 11$, this is used to make the D.O.T.S. method easier.

**Step 6.** Factor using the D.O.T.S. method to get your final answer.

$= (x+3+\sqrt{11})(x+3-\sqrt{11})$

E.g. 2. Factor $3x^2+12x-30$

**Step 1.** Take out the common factor if there is one.

$3(x^2+4x-10)$, has a common factor of 3

**Step 2.** Check to see if the factoring quadratic trinomial method works.

$3(x^2+4x-10)$, cannot be factored with the quadratic trinomial method.

**Step 3.** You must create a perfect square. You do this by:

i) halving the coefficient of the middle term $4 \div 2 = 2$

ii) then square this number $2^2 = 4$

iii) then add **and** subtract this number to the expression as the third and fourth terms as you did in E.g. 1. above.

$3(x^2 + 4x + 4 - 4 - 10)$

**Step 4.** Put parentheses around the perfect square and work out the last two digits.

$= 3[(x^2 + 4x + 4) - 4 - 10]$

$= 3[(x^2 + 4x + 4) - 14]$

**Step 5.** Factor the Perfect Square, and you will have a D.O.T.S.

$= 3[(x+2)^2 - 14]$

Note that $\sqrt{(14)^2} = 14$, this is used to make the D.O.T.S. method easier.

**Step 6.** Factor using the D.O.T.S. method to get your final answer.

$= 3(x + 2 + \sqrt{14})(x + 2 - \sqrt{14})$

E.g. 3. Factor $x^2 + 9x + 17$

Sometimes the middle term is on odd number. This is where you will need fractions to help you. Fractions keeps the calculations simpler. Using decimals can complicate the work.

**Step 1.** Take out the common factor if there is one.

$x^2 + 9x + 17$, has no common factor

**Step 2.** Check to see if the factoring quadratic trinomial method works.

$x^2 + 9x + 17$, cannot be factored with the quadratic trinomial method.

**Step 3.** You must create a perfect square. You do this by:

i) halving the coefficient of the middle term $9 \div 2 = \dfrac{9}{2}$

ii) then square this number $(\dfrac{9}{2})^2 = \dfrac{81}{4}$

iii) then add **and** subtract this number to the expression as the third and fourth terms as you've done earlier.

$x^2 + 9x + \dfrac{81}{4} - \dfrac{81}{4} + 17)$

**Step 4.** Put parentheses around the perfect square and work out the last two digits.

$(x^2 + 9x + \dfrac{81}{4}) - \dfrac{81}{4} + 17)$

$(x^2 + 9x + \dfrac{81}{4}) - \dfrac{13}{4}$

**Step 5.** Factor the Perfect Square, and you will have a D.O.T.S.

$= (x + \dfrac{9}{2})^2 - \dfrac{13}{4}$

Note that $\dfrac{13}{4} = \sqrt{(\dfrac{13}{4})^2}$, this is used to make the D.O.T.S. method easier.

**Step 6.** Factor using the D.O.T.S. method to get your final answer.

$$= (x + \dfrac{9}{2} + \dfrac{\sqrt{13}}{2})(x + \dfrac{9}{2} - \dfrac{\sqrt{13}}{2})$$

If you are unsure of your fractions, you can obtain the Easy Steps Math Fractions book from Amazon.com.

# Chapter 10

**Algebraic Fractions**

Algebraic fractions are algebraic expressions in fraction form. Algebraic fractions follow all the same rules as numerical fractions.

## Multiplication and Division of Algebraic Fractions

Examples.

E.g. 1. Simplify $\dfrac{4x}{10}$

**Step 1.** Look for any common factors in the numerator and the denominator.

This will be 2

**Step 2.** Cancel this factor.

$$\dfrac{\overset{2}{\cancel{4}}x}{\underset{5}{\cancel{10}}}$$

$= \dfrac{2x}{5}$ is the final answer

E.g. 2. Simplify $\dfrac{(x+2)(x+3)}{(x-4)(x+2)}$

**Step 1.** Look for any common factors in the numerator and the denominator.

This will be $(x+2)$

**Step 2.** Cancel this factor.

$$\frac{\overset{1}{\cancel{(x+2)}}(x+3)}{(x-4)\cancel{(x+2)}^1}$$

$$= \frac{x+3}{x-4} \text{ is the final answer}$$

E.g. 3. Simplify $\dfrac{(x+2)(x+5)}{(x-4)(x+4)} \times \dfrac{(x+2)(x+4)}{(x-5)(x+2)}$

**Step 1.** Look for any common factors in the numerators and the denominators.

These will be $(x+2)$ and $(x+4)$

**Step 2.** Cancel these factors.

$$\frac{\overset{1}{\cancel{(x+2)}}(x+5)}{(x-4)\cancel{(x+4)}^1} \times \frac{(x+2)\cancel{(x+4)}^1}{(x-5)\cancel{(x+2)}^1}$$

**Step 3.** Multiply across the top and multiply across the bottom

$$= \frac{(x+5)(x+2)}{(x-4)(x-5)} \text{ is the final answer}$$

E.g. 4. $\dfrac{(5x-3)(2x+3)}{(x-4)(x+7)} \div \dfrac{(2x+3)(5x-4)}{(x-3)(x+7)}$

This question is a division of fractions.

**Step 1.** Invert the second fraction and change the operation to a times (×)

$$\dfrac{(5x-3)(2x+3)}{(x-4)(x+7)} \times \dfrac{(x-3)(x+7)}{(2x+3)(5x-4)}$$

**Step 2.** Look for any common factors in the numerators and the denominators.

These will be $(2x+3)$ and $(x+7)$

**Step 3.** Cancel these factors.

$$\dfrac{(5x-3)\cancel{(2x+3)}^{1}}{(x-4)\cancel{(x+7)}^{1}} \times \dfrac{(x-3)\cancel{(x+7)}^{1}}{\cancel{(2x+3)}^{1}(5x-4)}$$

**Step 4.** Multiply across the top and multiply across the bottom

$= \dfrac{(5x-3)(x-3)}{(x-4)(5x-4)}$ is the final answer

E.g. 5. Simplify $\dfrac{4x^2-12x}{x^2-25} \times \dfrac{x^2+7x+10}{2x^2+4x}$

Sometimes algebraic expressions are not factored. These will need to be factored before they are simplified.

**Step 1.** Factor all the expressions in the fractions.

$$\dfrac{4x(x-3)}{(x+5)(x-5)} \times \dfrac{(x+2)(x+5)}{2x(x+2)}$$

**Step 2.** Look for common factors in the numerators and the denominators

These will be $(x+2)$, $(x+5)$, $x$ and $2$

**Step 3.** Cancel these factors

$$= \dfrac{\overset{2}{\cancel{4}}\cancel{x}(x-3)}{\cancel{(x+5)}(x-5)} \times \dfrac{\overset{1}{\cancel{(x+2)}}\overset{1}{\cancel{(x+5)}}}{\underset{1}{\cancel{2}}\cancel{x}\,\underset{1}{\cancel{(x+2)}}}$$

**Step 4.** Multiply across the top and multiply across the bottom

$= \dfrac{2(x-3)}{x-5}$ is the final answer

E.g. 6. $\dfrac{2x^2+5x-12}{3x^2+12x} \div \dfrac{4x^2-9}{12x}$

Again this is a division of fractions.

**Step 1.** Invert the second fraction and change the operation to a times (×).

$$\dfrac{2x^2+5x-12}{3x^2+12x} \times \dfrac{12x}{4x^2-9}$$

**Step 2.** Factor the algebraic expressions

$$= \dfrac{(2x-3)(x+4)}{3x(x+4)} \times \dfrac{12x}{(2x+3)(2x-3)}$$

**Step 3.** Look for common factors in the numerators and the denominators

These will be $(x+4)$, $(2x-3)$, $x$ and $3$

**Step 4.** Cancel these factors

$$= \dfrac{\cancel{(2x-3)}^{1}\, \cancel{(x+4)}^{1}}{\cancel{3x}^{1}\, \cancel{(x+4)}^{1}} \times \dfrac{\cancel{12x}^{4}}{(2x+3)\cancel{(2x-3)}^{1}}$$

**Step 5.** Multiply across the top and multiply across the bottom

$$= \dfrac{4}{2x+3} \text{ is the final answer}$$

**Simplify these Multiplications and Divisions.**

Remember to factor the expression where needed.

a) $\dfrac{18a(a-4)}{4a}$

b) $\dfrac{(x+1)(x+6)}{(x-5)(x+1)}$

c) $\dfrac{7x(x-4)}{10x} \div \dfrac{2(x-4)}{5}$

d) $\dfrac{(x+1)(x-3)}{(x-4)(x+7)} \times \dfrac{(x+7)(x+2)}{(x-1)(x+1)}$

e) $\dfrac{(2x+3)(5x-3)}{(x-4)(x+7)} \div \dfrac{(2x+3)(5x-4)}{(x-3)(x+7)}$

f) $\dfrac{x^2+11x+24}{x+3}$

g) $\dfrac{3x+6}{4x-24} \times \dfrac{8x-48}{6x-12}$

h) $\dfrac{8x^2-2x-3}{4x^2+3x-10} \times \dfrac{16x^2-25}{6x^2+7x+2}$

i) $\dfrac{x^2-4}{x^2-9} \div \dfrac{4x+8}{12x-36}$

j) $\dfrac{x^2-2x+1-4y^2}{12x^2-5x-2} \div \dfrac{3x-3+6y}{28x+7}$

## Addition and Subtraction of Algebraic Fractions

The rules that apply to addition and subtraction of fractions also apply to algebraic fractions.

Examples

E.g. 1. Simplify $\dfrac{a}{3}+\dfrac{a}{4}$

**Step 1.** Find the lowest common denominator of these fractions.

LCD of 3 and 4 is 12

**Step 2.** Rewrite the fractions as equivalent fractions using the LCD.

$= \dfrac{4a}{12}+\dfrac{3a}{12}$

The two numerators can now be added, or the two fractions can be written as a single fraction

**Step 3.** Rewrite the expression as a single fraction

$= \dfrac{4a+3a}{12}$

**Step 4.** Collect like terms in the numerator

$= \dfrac{7a}{12}$ this is the final answer.

E.g. 2. Simplify $\dfrac{a+5}{4} - \dfrac{a-2}{5}$

**Step 1.** Find the lowest common denominator of these fractions.

LCD of 4 and 5 is 20
**Step 2.** Rewrite the fractions as equivalent fractions using the LCD.

$= \dfrac{5(a+5)}{20} - \dfrac{4(a-2)}{20}$

The two fractions can now be written as a single fraction

**Step 3.** Rewrite the expression as a single fraction

$= \dfrac{5(a+5)-4(a-2)}{20}$

**Step 4.** Expand the brackets in the numerator. Be careful of the negative when expanding.

$= \dfrac{5a+25-4a+8}{20}$

**Step 5.** Collect the like terms in the numerator

$= \dfrac{a+33}{20}$ this is the final answer.

E.g. 3. $\dfrac{3}{x+2} + \dfrac{5}{x-4}$

**Step 1.** Find the lowest common denominator of these fractions.

LCD of $x+2$ and $x-4$ is $(x+2)(x-4)$

**Step 2.** Rewrite the fractions as equivalent fractions using the LCD.

$$= \dfrac{3(x-4)}{(x+2)(x-4)} + \dfrac{5(x+2)}{(x+2)(x-4)}$$

The two fractions can now be written as a single fraction

**Step 3.** Rewrite the expression as a single fraction

$$= \dfrac{3(x-4)+5(x+2)}{(x+2)(x-4)}$$

**Step 4.** Expand the brackets in the numerator

$$= \dfrac{3x-12+5x+10}{(x+2)(x-4)}$$

**Step 5.** Collect like terms in the numerator

$$= \dfrac{8x-2}{(x+2)(x-4)} \text{ this is the final answer.}$$

E.g. 4. Simplify $\dfrac{2}{(x+3)(x-1)} + \dfrac{3}{(x-2)(x+3)}$

**Step 1.** Find the lowest common denominator of these fractions.

LCD of $(x+3)(x-1)$ and $(x-2)(x+3)$ is $(x+3)(x-1)(x-2)(x+3)$. Notice that the expression $(x+3)$ appears twice.

**Step 2.** Rewrite the fractions as equivalent fractions using the LCD.

$$= \dfrac{2(x-2)(x+3)}{(x+3)(x-1)(x-2)(x+3)} + \dfrac{3(x+3)(x-1)}{(x+3)(x-1)(x-2)(x+3)}$$

Note that $(x+3)$ appears in the numerator and twice in the denominator of both fractions. You can cancel one of these from each numerator and one from each denominator.

$$= \dfrac{2(x-2)\cancel{(x+3)}}{\cancel{(x+3)}(x-1)(x-2)(x+3)} + \dfrac{3\cancel{(x+3)}(x-1)}{\cancel{(x+3)}(x-1)(x-2)(x+3)}$$

$$= \dfrac{2(x-2)}{(x-1)(x-2)(x+3)} + \dfrac{3(x-1)}{(x-1)(x-2)(x+3)}$$

The two fractions can now be written as a single fraction

**Step 3.** Rewrite the expression as a single fraction

$$= \dfrac{2(x-2)+3(x-1)}{(x-1)(x-2)(x+3)}$$

**Step 4.** Expand the brackets in the numerator

$$= \frac{2x-4+3x-3}{(x-1)(x-2)(x+3)}$$

**Step 5.** Collect like terms in the numerator

$$= \frac{5x-7}{(x-1)(x-2)(x+3)} \quad \text{this is the final answer.}$$

**Simplify these Additions and Subtractions**

a) $\dfrac{x}{2} + \dfrac{x}{5}$

b) $\dfrac{3x-8}{4} - \dfrac{x-3}{10}$

c) $\dfrac{7}{5x} + \dfrac{2}{15x}$

d) $\dfrac{2}{7} - \dfrac{1}{3x}$

e) $\dfrac{5}{x+7} - \dfrac{2}{x-6}$

f) $\dfrac{5}{2(x-2)} + \dfrac{2}{2-x}$

g) $\dfrac{8}{(7x-3)^2} - \dfrac{5}{7x-3}$

h) $\dfrac{1}{(x+3)(x+4)} + \dfrac{5}{(x+4)(x+7)}$

i) $\dfrac{7}{(3x-2)(x+5)} - \dfrac{2}{(3x-2)^2}$

j) $\dfrac{3}{x^2+2x-15} - \dfrac{6}{x^2+9x+20}$

# Chapter 11

## Solving Linear Equations

**Linear relationships** are relationships between **two** variables that produce a straight line when graphed on a Cartesian plane, like the one below.

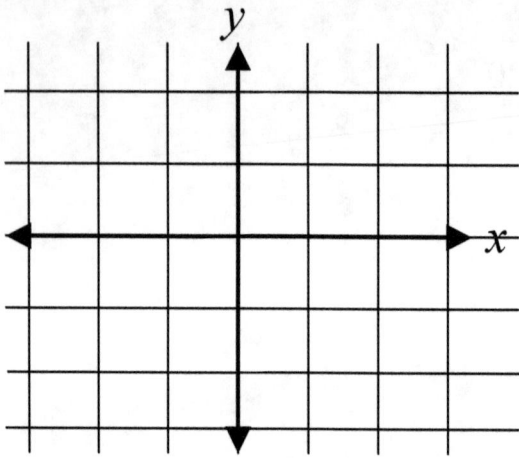

A linear relationship does not have any exponents or square roots or cube roots etc.

Examples of linear relations are:

$3x+2y=4$

$y=3x+4$

Note that there are two unknown variables.

A **Linear equation**, on the other hand, is a linear relationship that already has a value for one of its variables. So to solve a linear equation, only one variable needs to be found.

For example the equation $3x-2=12$ has only one variable.
To solve a linear equation, you must first know that in algebra *to solve*, means to work out the value of the unknown variable. So you must get this variable by itself on one side of the equals sign and everything else on the other side.

Second, you must be aware of **inverse operations** and how and when to use them. Inverse operations refer to operations that have the opposite effect to each other. Inverse operations, refer to Plus $(+)$ and Minus $(-)$, Times $(\times)$ and Divide $(\div)$, Square $(x^2)$ and Square Root $(\sqrt{x})$.

*Plus $(+)$ and Minus $(-)$ are opposite operations, so they are inverse to each other.*
*Times $(\times)$ and Divide $(\div)$ are opposite operations, so they are inverse to each other.*
*Square $(x^2)$ and Square Root $(\sqrt{x})$ are opposite operations, so they are inverse to each other.*

Third, you must be aware that an equation is like a balanced scale with the equals sign in the middle (one side of the equation equals the other side of the equation). Whatever you do to the left of the equals sign, you **must** do to the right of the equals sign. This way the equation remains balanced.

And fourth, when solving algebraic expressions you reverse the process for PEMDAS or BODMAS.

Therefore, to solve an equation, you must get the unknown variable on its own to one side of the equals sign and everything else to the other side, and you do this by using inverse operations and reversing PEMDAS.

Examples

E.g. 1. Solve $5a-6=34$

**Step 1.** Reversing PEMDAS (BODMAS), use inverse operations to 'undo' any additions or subtractions near the variable. You must balance the equation by doing the same on the other side of the equals sign.

$5a-6+6=34+6$

$5a=40$

**Step 2.** Again use inverse operations to 'undo' any multiplications or divisions.

$$\frac{5a}{5}=\frac{40}{5}$$

**Step 3.** Simplify terms by canceling.

$$\frac{\cancel{5}a}{\cancel{5}}=\frac{\cancel{40}^{8}}{\cancel{5}}$$

$a=8$ is your final answer.

E.g. 2. Solve $\dfrac{3x}{2}+4=-1$

**Step 1.** Reversing PEMDAS, use inverse operations to 'undo' any additions or subtractions near the variable. Do the same on the other side of the equals sign.

$$\frac{3x}{2}+4-4=-1-4$$

$$\frac{3x}{2}=-5$$

**Step 2.** Again use inverse operations to 'undo' any multiplications or divisions. It is usually easier to reverse the division first if both need to be completed. Remember to cancel.

$$\frac{3x}{2}\times 2=-5\times 2$$

$$\frac{3x}{\cancel{2}}\times \cancel{2}=-5\times 2$$

$$3x=-10$$

**Step 3.** Again use inverse operations to 'undo' any remaining multiplications or divisions.

$$\frac{3x}{3}=\frac{-10}{3}$$

**Step 4.** Simplify terms by canceling.

$$\frac{\cancel{3}x}{\cancel{3}}=\frac{-10}{3}$$

$$x=\frac{-10}{3}=-3\frac{1}{3}$$ is your final answer.

E.g. 3. Solve $2(2d-5)=6$

**Step 1.** Reversing PEMDAS, use inverse operations to 'undo' the multiplication before parentheses. Remember to do the same thing on both sides of the equals sign. (If you choose to expand the parentheses first before solving, you will get the same answer).

$$\frac{2(2d-5)}{2}=\frac{6}{2}$$

$$2d-5=3$$

**Step 2.** Use inverse operations to 'undo' any additions or subtractions near the variable.

$$2d-5+5=3+5$$

$$2d=8$$

**Step 3.** Use inverse operations to 'undo' the multiplication

$$\frac{\cancel{2}d}{\cancel{2}}=\frac{\cancel{8}^{4}}{\cancel{2}}$$

$d=4$, is your final answer.


Sometimes the same variable can appear on both sides of the equals sign. Just follow the same rules for inverse operations to get them both to the same side, and then collect like terms.

E.g. 4. Solve $5x-3=2x+6$

**Step 1.** Use inverse operations to 'move' the variable from the right hand side (RHS) of the equals sign to the left hand side (LHS).

$$5x-3-2x=2x+6-2x$$

$$3x-3=6$$

**Step 2.** Now use inverse operations to 'undo' the subtraction (or to 'move it from the LHS to the RHS).

$$3x-3+3=6+3$$

$$3x=9$$

**Step 3.** Use inverse operations to 'undo' the multiplication

$$\frac{3x}{3}=\frac{9}{3}$$

$$\frac{\cancel{3}x}{\cancel{3}}=\frac{\cancel{9}^3}{\cancel{3}}$$

$x=3$ is your final answer

E.g. 5. Solve $4(x+1)=3(6-x)$

**Step 1.** Expand the parentheses.

$$4x+4=18-3x$$

**Step 2.** Use inverse operations to 'move' the variable from the RHS of the equals sign to the left.

$$4x+4+3x=18-3x+3x$$

$$7x+4=18$$

**Step 3.** Now use inverse operations to 'undo' the addition.

$$7x+4-4=18-4$$

$$7x=14$$

**Step 4.** Use inverse operations to 'undo' the multiplication.

$$\frac{7x}{7}=\frac{14}{7}$$

$$\frac{\cancel{7}x}{\cancel{7}}=\frac{\cancel{14}^2}{\cancel{7}}$$

$x=2$ is your final answer

E.g. 6. Solve $\dfrac{2n+1}{3}=\dfrac{n-3}{5}$

When you have two fractions that equal each other like in the E.g. 6 above, you can use a technique called **cross multiplication**. Cross multiplication allows you to multiply the numerator of the LHS fraction with the denominator of the RHS fraction and the numerator of the RHS fraction with the denominator of the LHS fraction. By doing this, you end up with an equation that looks like the one in E.g. 5 (don't forget the parentheses).

That is,

$$\frac{2n+1}{3} \diagup\!\!\!\diagdown \frac{n-3}{5}$$

Therefore:

**Step 1.** Cross multiply

$$\frac{2n+1}{3} \diagup\!\!\!\diagdown \frac{n-3}{5}$$

$5(2n+1) = 3(n-3)$

**Step 2.** Expand the parentheses

$10n + 5 = 3n - 9$

**Step 3.** Use inverse operations to 'move' the variable from the RHS of the equals sign to the left.

$10n + 5 - 3n = 3n - 9 - 3n$

$7n + 5 = -9$

**Step 4.** Now use inverse operations to 'undo' the addition on the RHS.

$7n + 5 - 5 = -9 - 5$

$7n = -14$

**Step 5.** Use inverse operations to 'undo' the multiplication

$$\frac{7n}{7} = -\frac{14}{7}$$

$$\frac{\cancel{7}n}{\cancel{7}} = -\frac{\cancel{14}^{2}}{\cancel{7}}$$

$n = -2$ is your final answer

E.g. 7. Solve $\dfrac{2a+3}{4} - \dfrac{3a-5}{3} = 2$

This type of question requires you to use your knowledge of fractions to solve it. Cross multiplication will not work, as the fractions are not equal to each other.

Step 1. Make all denominators the same by finding the LCM for all three terms.

$\dfrac{3(2a+3)}{12} - \dfrac{4(3a-5)}{12} = \dfrac{24}{12}$ LCM is 12

Step 2. Rewrite the expression without the denominators. In algebra if all terms have the same denominator, then the expression can be written without the denominators and this doesn't affect the answer.

$3(2a+3) - 4(3a-5) = 24$

Step 3. Expand the brackets and collect like terms

$6a + 9 - 12a + 20 = 24$

$-6a + 29 = 24$

Step 4. Use inverse operations to 'undo' the addition

$-6a + 29 - 29 = 24 - 29$

$-6a = -5$

Step 5. Use inverse operations to 'undo' the multiplication

$$\frac{-6a}{-6} = \frac{-5}{-6}$$

$$\frac{\cancel{-6}a}{\cancel{-6}} = \frac{\cancel{-}5}{\cancel{-}6}$$

$$a = \frac{5}{6}$$

**Solve the following Equations**

a) $4x - 4 = 16$

b) $-2 - 3a = 4$

c) $\frac{5a}{3} - 4 = 6$

d) $3 - \frac{4b}{2} = 5$

e) $3x + 7 = x + 9$

f) $-6 - r = 2r + 1$

g) $7(c + 2) = 2(2c - 1)$

h) $2(4m + 3) = 3(1 - m)$

i) $\frac{x+7}{3} = \frac{x-3}{5}$

j) $\frac{4e-3}{7} + \frac{2e+5}{2} = 3$

Sometimes you will be given a problem to solve, however you will need to write an equation in order to solve it.

E.g. The sum of three consecutive numbers is 48. What are the numbers?
Firstly you need to know what the parts of the question mean.

**Sum** means addition, and **consecutively** means one after another.

So you need to find three numbers that are one after the other that add up to 48.

Step 1. Start with $x$. This is your first number.

Step 2. Find the next two numbers

The number right after $x$ must be $x+1$. The number after $x+1$ must be $x+2$.

Step 3. Add these three numbers and make it equal 48.

$x+x+1+x+2=48$

Step 4. Collect like terms

$3x+3=48$

Step 5. Solve the equation

$3x=45$

$x=15$

Therefore the three numbers are 15, 16 and 17.

**Solve these worded problems after writing an equation.**

a) 6 is added to $x$ to give the result of 13.

b) 7 is subtracted from $a$ to give a result of $-2$.

c) Twice a certain number is 18.
d) One-third of a certain number is 24.

e) Two more than three times a number is 23

f) Five less than four times a number is $-3$

g) The sum of four consecutive numbers is 38. What are the numbers?
h) The perimeter of a square is 100 cm. What is the length of each side?

i) Find three consecutive even numbers whose sum is 126.

j) Five is subtracted from three times a number to give 25. What is the number?

# **Multiplication Tables**

To make calculations really easy, learn your multiplications tables. Here is a set of multiplication tables from 1 x 1 to 12 x 12 to help you if you need it.

| | | | |
|---|---|---|---|
| 1 x 1 = 1   | 2 x 1 = 2   | 3 x 1 = 3   | 4 x 1 = 4   |
| 1 x 2 = 2   | 2 x 2 = 4   | 3 x 2 = 6   | 4 x 2 = 8   |
| 1 x 3 = 3   | 2 x 3 = 6   | 3 x 3 = 9   | 4 x 3 = 12  |
| 1 x 4 = 4   | 2 x 4 = 8   | 3 x 4 = 12  | 4 x 4 = 16  |
| 1 x 5 = 5   | 2 x 5 = 10  | 3 x 5 = 15  | 4 x 5 = 20  |
| 1 x 6 = 6   | 2 x 6 = 12  | 3 x 6 = 18  | 4 x 6 = 24  |
| 1 x 7 = 7   | 2 x 7 = 14  | 3 x 7 = 21  | 4 x 7 = 28  |
| 1 x 8 = 8   | 2 x 8 = 16  | 3 x 8 = 24  | 4 x 8 = 32  |
| 1 x 9 = 9   | 2 x 9 = 18  | 3 x 9 = 27  | 4 x 9 = 36  |
| 1 x 10 = 10 | 2 x 10 = 20 | 3 x 10 = 30 | 4 x 10 = 40 |
| 1 x 11 = 11 | 2 x 11 = 22 | 3 x 11 = 33 | 4 x 11 = 44 |
| 1 x 12 = 12 | 2 x 12 = 24 | 3 x 12 = 36 | 4 x 12 = 48 |

| | | | |
|---|---|---|---|
| 5 x 1 = 5 | 6 x 1 = 6 | 7 x 1 = 7 | 8 x 1 = 8 |
| 5 x 2 = 10 | 6 x 2 = 12 | 7 x 2 = 14 | 8 x 2 = 16 |
| 5 x 3 = 15 | 6 x 3 = 18 | 7 x 3 = 21 | 8 x 3 = 24 |
| 5 x 4 = 20 | 6 x 4 = 24 | 7 x 4 = 28 | 8 x 4 = 32 |
| 5 x 5 = 25 | 6 x 5 = 30 | 7 x 5 = 35 | 8 x 5 = 40 |
| 5 x 6 = 30 | 6 x 6 = 36 | 7 x 6 = 42 | 8 x 6 = 48 |
| 5 x 7 = 35 | 6 x 7 = 42 | 7 x 7 = 49 | 8 x 7 = 56 |
| 5 x 8 = 40 | 6 x 8 = 48 | 7 x 8 = 56 | 8 x 8 = 64 |
| 5 x 9 = 45 | 6 x 9 = 54 | 7 x 9 = 63 | 8 x 9 = 72 |
| 5 x 10 = 50 | 6 x 10 = 60 | 7 x 10 = 70 | 8 x 10 = 80 |
| 5 x 11 = 55 | 6 x 11 = 66 | 7 x 11 = 77 | 8 x 11 = 88 |
| 5 x 12 = 60 | 6 x 12 = 72 | 7 x 12 = 84 | 8 x 12 = 96 |

| | | | |
|---|---|---|---|
| 9 x 1 = 9 | 10 x 1 = 10 | 11 x 1 = 11 | 12 x 1 = 12 |
| 9 x 2 = 18 | 10 x 2 = 20 | 11 x 2 = 22 | 12 x 2 = 24 |
| 9 x 3 = 27 | 10 x 3 = 30 | 11 x 3 = 33 | 12 x 3 = 36 |
| 9 x 4 = 35 | 10 x 4 = 40 | 11 x 4 = 44 | 12 x 4 = 48 |
| 9 x 5 = 45 | 10 x 5 = 50 | 11 x 5 = 55 | 12 x 5 = 60 |
| 9 x 6 = 54 | 10 x 6 = 60 | 11 x 6 = 66 | 12 x 6 = 72 |
| 9 x 7 = 63 | 10 x 7 = 70 | 11 x 7 = 77 | 12 x 7 = 84 |
| 9 x 8 = 72 | 10 x 8 = 80 | 11 x 8 = 88 | 12 x 8 = 96 |
| 9 x 9 = 81 | 10 x 9 = 90 | 11 x 9 = 99 | 12 x 9 = 108 |
| 9 x 10 = 90 | 10 x 10 = 100 | 11 x 10 =110 | 12 x 10 = 120 |
| 9 x 11 = 99 | 10 x 11 = 110 | 11 x 11 = 121 | 12 x 11 = 132 |
| 9 x 12 = 108 | 10 x 12 = 120 | 11 x 12 = 132 | 12 x 12 = 144 |

# Answers

Rewrite without signs

a) $\dfrac{x}{6}$  b) $\dfrac{3}{y}$  c) $4a$  d) $7b$  e) $\dfrac{5}{2a}$  f) $\dfrac{9}{a}-\dfrac{x}{7}$  g) $\dfrac{ab}{c}+3a$

h) $\dfrac{9}{3s}+\dfrac{6r}{uv}$  i) $\dfrac{4ef}{qt}$  j) $\dfrac{sv}{7}+15$

Identify the Parts

a) i) $a$   ii) 1   iii) 1   iv) $b$   v) 0

b) i) $a$   ii) –1   iii) 1   iv) $-b$   v) –2

c) i) $a$   ii) 2   iii) 1   iv) $2b$   v) 0

d) i) $x$   ii) 1   iii) 1   iv) $y$   v) 3

e) i) $x$   ii) 4   iii) 1   iv) $4y$   v) 2

f) i) $x$   ii) 3   iii) 1   iv) $3y$   v) 0

g) i) $x$   ii) 2   iii) 2   iv) $2y$   v) 0

h) i) $x$   ii) 3   iii) 2   iv) $3y$   v) 2

i) i) $x$   ii) 0   iii) 2   iv) 25   v) 25

j) i) $xy$   ii) 1   iii) 2   iv) $\dfrac{x}{4}$   v) –2

Simplify the Following Expressions (Like Terms)

a) $17a$  b) $7f$  c) $-4w$  d) $13ef^2g^3$  e) $12d-10e$
f) $16ij+3j$  g) $10x^3+9y^2$  h) $-8x^2-10y+2$
i) $3b^3-b-5ab$  j) $150stc+40st+9s-5c+20$

Complete the Substitutions

a) 23  b) 2  c) $-8$  d) 2  e) 36  f) $-17$  g) $-29$  h) $-4$  i) 12  j) $-7½$

Complete the Substitutions with Exponents

a) 9  b) 20  c) 108  d) 16  e) $-63$  f) 78  g) 2  h) $-3$  i) 1  j) 39

Multiplying Algebraic Terms

a) $24y$  b) $27xy$  c) $-14ab$  d) $10a$  e) $44tv$  f) $36st^2$
g) $18a^2$  h) $30x^2y$  i) $24x^2y^2$  j) $80def$

Dividing Algebraic Terms

a) $a^5$  b) $9x^2$  c) $2x^5$  d) $\dfrac{3}{x^4}$  e) $\dfrac{x^6}{3}$  f) $\dfrac{7xy^3}{5}$  g) $-3r$

h) $-\dfrac{9}{4a^3bc}$  i) $-\dfrac{1}{8}$  j) $\dfrac{4x^4}{3y^3z^5}$

Expand and Simplify a(b+c) type

a) $4x+40$  b) $3bc-3bd$  c) $6x+8$  d) $5x^2+10xy$
e) $-12a^3b+6a^3$  f) $8x^7y+24x^5y^4$  g) $10d+3de$
h) $-7q-56qr$  i) $8a^5-8a^2+8a^6-a$  j) $-21x^5-6x^2+12x^7$

Expand and Simplify (a+b)(c+d) type

a) $x^2+4x+3$  b) $x^2-x-20$  c) $3x^2+7x+2$
d) $4x^2-10x-6$  e) $3x^2-12$  f) $3x^2+7x+2$
g) $4x^2-10x-6$  h) $6x^2-11x-35$  i) $8x^2+50x+63$
j) $-18x^2+27x-10$

Expand and Simplify Perfect Square $(a+b)^2$ type

a) $x^2+2x+1$  b) $x^2-6x+9$  c) $x^2+14x+49$
d) $4x^2-8x+4$  e) $9x^2+24x+16$  f) $4x^2-16x+16$
g) $9x^2-30x+25$  h) $9x^2-12x+4$  i) $4x^2+24x+36$
j) $36y^2-60xy+25x^2$

Expand and Simplify Difference of Two Squares (a+b)(a-b) type

a) $x^2-9$  b) $x^2-81$  c) $49-x^2$  d) $9-36x^2$  e) $25x^2-16$
f) $64-49x^2$  g) $16-x^2$  h) $64-4x^2$  i) $16x^2-4y^2$
j) $4y^2-49x^2$

Expand and Simplify Three Factor Expressions

a) $x^3+9x^2+20x$  b) $x^3-4x^2+3x$  c) $-x^3+x^2+12x$
d) $2x^3-4x^2-96x$  e) $-3x^3+9x^2+3x$
f) $x^3-6x^2-4x+24$  g) $x^3-4x^2-7x+10$
h) $x^3-11x^2+36x-36$  i) $2x^3-x^2-25x-12$
j) $2x^3+12x^2+10x-24$

Factor the Binomial Expressions
a) $3(x-15)$  b) $4(5-x)$  c) $2(xy-9)$  d) $x(x-4)$
e) $4x(3y-x)$  f) $ax(7x-2a)$  g) $4ab(b-1)$  h) $15x(y+2)$
i) $10x(4y-3)$  j) $4y(8x+9)$

Factor using the D.O.T.S. method

a) $(b+2)(b-2)$  b) $(7x+yz)(7x-yz)$  c) $7(a+b)(a-b)$
d) $3(ab+2c)(ab-2c)$  e) $(x+5)(x+1)$  f) $3(3x-1)(1-x)$
g) $3(x-2)(x-6)$  h) $(x+\sqrt{3})(x-\sqrt{3})$  i) $3(3a+\sqrt{5})(3a-\sqrt{5})$
j) $(6x-11)(4x+3)$

Factor using the Perfect Square method

a) $(x+3)^2$  b) $(x+2)^2$  c) $(x-7)^2$  d) $(x-4)^2$
e) $(x+1)^2$  f) $(2x+5)^2$  g) $(3x+4)^2$  h) $(5x-1)^2$
i) $3(x+6)^2$  j) $2(5x-4)^2$

Factor using the Two and Two method

a) $(b+5)(a+3)$  b) $(3+z)(x+y)$  c) $(2j-3)(4v-3)$
d) $(b+2)(b+4)$  e) $(c-3)(c-4)$  f) $(3d+2)(4d+3)$
g) $(a-c)(a+c+7)$  h) $(4b+5e)(4b-5e+2)$
i) $(9-7m)(9+7m-3n)$  j) $(4-3n)(4+3n-4s)$

Factor using the Three and One method

a) $(x+4+y)(x+4-y)$  b) $(h-2+k)(h-2-k)$
c) $(a+4+b)(a+4-b)$  d) $(r-5+s)(r-5-s)$
e) $(b+1+3n)(b+1-3n)$  f) $(j+3+4t)(j+3-4t)$
g) $(x-3y+2)(x-3y-2)$  h) $(a-5b+6)(a-5b-6)$
i) $(2m-3+4n)(2m-3-4n)$  j) $(3v+5+2j)(3v+5-2j)$

Factor these Quadratic Trinomials

a) $(x+3)(x+8)$  b) $(x-5)(x+6)$  c) $(z-5)(z+12)$
d) $(2a+1)(a+2)$  e) $(3m-4)(m+4)$
f) $2(2j-1)(3j-1)$  g) $(2b+1)(2b+3)$
h) $(4s+3)(3s-5)$  i) $3(x+4)(x+2)$  j) $5(a+7)(a-3)$

Solve the Equations

a) $x=5$  b) $a=-2$  c) $a=6$  d) $b=-1$  e) $x=1$
f) $r=-\dfrac{7}{3}=-2\dfrac{1}{3}$  g) $c=-\dfrac{16}{3}=-5\dfrac{1}{3}$  h) $m=-\dfrac{3}{11}$
i) $x=-22$  j) $e=\dfrac{13}{22}$

Simplify Algebraic Terms – Multiplication & Division

a) $\dfrac{9(a-4)}{2}$  b) $\dfrac{x+6}{x-5}$  c) $\dfrac{7}{4}$  d) $\dfrac{(x-3)(x+2)}{(x-4)(x-1)}$

e) $\dfrac{(5x-3)(x-3)}{(x-4)(5x-4)}$  f) $x+8$  g) $\dfrac{x+2}{x-2}$  h) $\dfrac{(4x-3)(4x+5)}{(x+2)(3x+2)}$

i) $\dfrac{3(x-2)}{x+3}$  j) $\dfrac{7(x-1-2y)}{3(3x-2)}$

Simplify Algebraic Terms – Addition and Subtraction

a) $\dfrac{7x}{10}$  b) $\dfrac{13x-34}{20}$  c) $\dfrac{23}{15x}$  d) $\dfrac{6x-7}{21x}$  e) $\dfrac{3x-44}{(x+7)(x-6)}$

f) $\dfrac{1}{2(x-2)}$  g) $\dfrac{23-35x}{(7x-3)^2}$  h) $\dfrac{2(3x-11)}{(x+3)(x+4)(x+7)}$

i) $\dfrac{19x-24}{(x+5)(3x-2)^2}$  j) $\dfrac{3(10-x)}{(x-3)(x+4)(x+5)}$

Worded Problems

a) $x+6=13$, $x=7$  b) $a-7=-2$, $a=5$
c) $2x=18$, $x=9$  d) $\dfrac{x}{3}=24$, $x=72$
e) $3x+2=23$, $x=7$  f) $4x-5=-3$, $x=\dfrac{1}{2}$
g) $4x+6=38$, 8,9,10  h) $4l=100$, $l=25$
i) $3x+6=126$, 40,42,44  j) $3x-5=25$, $x=10$

# Glossary of Useful Terms

**Sum** refers to addition. The sum of two numbers is the answer of one number **plus** another number. E.g. the sum of 2 and 6 is 8, (2 + 6 = 8).

**Difference** refers to subtraction. The difference between two numbers is the answer of one number **minus** another number. E.g. the difference between 6 and 2 is 4, (6 − 2 = 4).

**Product** refers to multiplication. The product of two numbers is the answer of one number **times** another number. E.g. the product of 2 and 6 is 12, (2 x 6 = 12).

**Quotient** refers to division. The quotient is the answer of one number being **divided** by another number. E.g. the quotient of 6 and 2 is 3 (6 ÷ 2 = 3).

**Greatest Common Divisor** (GCD) or **Greatest Common Factor** (GCF), or **Highest Common Factor** (HCF): is the largest number or factor that goes into two other larger numbers without remainders.

E.g.

Q. *What is the GCD of 18 and 24?*

The factors of 24 are 1, 2, 3, 4, 6, 8, 12 and 24.

The factors of 18 are 1, 2, 3, 6, 9, and 18.

The largest number that is in both sets above is 6. Therefore the greatest common divisor of 24 and 18 is 6.

**Least Common Multiple** or **Lowest Common Multiple** (LCM): is the smallest number or multiple that two other numbers can both go in to.

E.G.

Q. What is the LCM of 3 and 4?

The multiples of 3 are 3, 6, 9, 12, 18, 21, etc

The multiples of 4 are 4, 8, 12, 16, 20, etc

The lowest number that is in both sets above is 12. Therefore the lowest common multiple of 3 and 4 is 12.

**Least Common Denominator** or **Lowest Common Denominator** (LCD): is the smallest denominator that two other denominators can both go in to. *(Used when adding and subtracting fractions)*

E.g.

Q. What is the LCD of $\frac{1}{2}$ and $\frac{1}{3}$?

The multiples of the denominator 2 are 2, 4, 6, 8, 10, etc

The multiples of the denominator 3 are 3, 6, 9, 12, etc

The lowest number that is in both sets above is 6. Therefore 6 will be used as the lowest common denominator.

$$\frac{1}{2} + \frac{1}{3} = \frac{-}{6} + \frac{-}{6}$$

Note that the working out for LCM and LCD is exactly the same.

To get the **Reciprocal** of a fraction, just turn the fraction upside down. E.g.

If the fraction is $\frac{2}{3}$, then it's reciprocal is $\frac{3}{2}$.

The reciprocal of 3 is $\frac{1}{3}$. The reciprocal of $\frac{1}{3}$ is 3.

The get the reciprocal of a mixed number like $2\frac{1}{2}$, first change it to an improper fraction $\frac{5}{2}$, and then turn it upside down $\frac{2}{5}$. So the reciprocal of $2\frac{1}{2}$ is $\frac{2}{5}$.

Any fraction multiplied by it's reciprocal always equals 1.

i.e. $\frac{2}{5} \times \frac{5}{2} = 1$

In a division, there are three words that should be learned. These are:
**Dividend** – the number being divided.
**Divisor** – the number doing the dividing?
**Quotient** – the answer.

This information can be shown with the division symbols.

$$Dividend \div Divisor = Quotient$$

or in a 'division box'

$$Divisor \overline{)Dividend}^{Quotient}$$

Or, as a fraction.

The <u>numerator is the dividend</u>, the <u>denominator is the divisor</u> and the <u>quotient is the answer</u>.

$$\frac{Dividend}{Divisor} = Quotient$$

**Reciprocal** – refers to a number or variable being inverted and written under the number 1.

For example the reciprocal of 5 is $\frac{1}{5}$, the reciprocal of $\frac{3}{4}$ is $\frac{4}{3}$, the reciprocal of $x$ is $\frac{1}{x}$.

**Surd** – refers to any number that is not a perfect square, but is written under the square root sign anyway. Surds can be converted to decimals, but for ease of calculations, they are left as they are. Examples of surds are $\sqrt{5}$ or $\sqrt{11}$ or $\sqrt{23}$ etc

Whenever you multiply a number by itself, the result is a **Perfect Square**. Here is a list of the first twelve numerical perfect squares as well as a couple of algebraic ones.

$1 \times 1 = 1^2 = 1$

$2 \times 2 = 2^2 = 4$

$3 \times 3 = 3^2 = 9$

$4 \times 4 = 4^2 = 16$

$5 \times 5 = 5^2 = 25$

$6 \times 6 = 6^2 = 36$

$7 \times 7 = 7^2 = 49$

$8 \times 8 = 8^2 = 64$

$9 \times 9 = 9^2 = 81$

$10 \times 10 = 10^2 = 100$

$11 \times 11 = 11^2 = 121$

$12 \times 12 = 12^2 = 144$

$x \times x = x^2$

$(x+1) \times (x+1) = (x+1)^2$

The square root of a perfect square is the number that was multiplied by itself.

$\sqrt{1}=1$

$\sqrt{4}=2$

$\sqrt{9}=3$

$\sqrt{16}=4$

$\sqrt{25}=5$

$\sqrt{36}=6$

$\sqrt{49}=7$

$\sqrt{64}=8$

$\sqrt{81}=9$

$\sqrt{100}=10$

$\sqrt{121}=11$

$\sqrt{144}=12$

$\sqrt{x^2}=x$ or $(\sqrt{x})^2=x$

$\sqrt{(x+1)^2}=(x+1)$

**Check out the other books in the Easy Steps Math series**

Fractions
Decimals
Percentages
Ratios
Negative Numbers
Master Collection 1 – Fractions, Decimals and Percentages
Master Collection 2 – Fractions, Decimals and Ratios
Master Collection 3 – Fractions, Percentages and Ratios
Master Collection 4 – Decimals, Percentages and Ratios

www.ingramcontent.com/pod-product-compliance
Lightning Source LLC
Chambersburg PA
CBHW051724170526
45167CB00002B/796